The Morality of Urban Mobility

Philosophy, Technology, and Society

***Series Editor*:** Sven Ove Hansson

Technological change has deep and often unexpected impacts on our societies. Sometimes new technologies liberate us and improve our quality of life, sometimes they bring severe social and environmental problems, sometimes they do both. This book series reflects philosophically on what new and emerging technologies do to our lives and how we can use them more wisely. It provides new insights on how technology continuously changes the basic conditions of human existence: relationships among us, our relations to nature, the knowledge we can obtain, our thought patterns, our ethical difficulties, and our views of the world.

Titles in the Series

The Ethics of Technology: Methods and Approaches, edited by Sven Ove Hansson
Nanotechnology: Regulation and Public Discourse, edited by Iris Eisenberger, Angela Kallhoff and Claudia Schwarz-Plaschg
Water Ethics: An Introduction, Neelke Doorn
Humans and Robots: Ethics, Agency, and Anthropomorphism, Sven Nyholm
Interpreting Technology: Ricoeur on Questions Concerning Ethics and Philosophy of Technology, edited by Wessel Reijers, Alberto Romele, and Mark Coeckelbergh
The Morality of Urban Mobility: Technology and Philosophy of the City, Shane Epting
Test-Driving the Future: Autonomous Vehicles and the Ethics of Technological Change, edited by Diane Michelfelder (forthcoming)
The Ethics of Behaviour Change Technologies, edited by Joel Anderson, Lily Frank, and Andreas Spahn (forthcoming)

The Morality of Urban Mobility

Technology and Philosophy of the City

Shane Epting

Foreword by
Lewis Gordon

ROWMAN & LITTLEFIELD
Lanham • Boulder • New York • London

Published by Rowman & Littlefield
An imprint of The Rowman & Littlefield Publishing Group, Inc.
4501 Forbes Boulevard, Suite 200, Lanham, Maryland 20706
www.rowman.com

86-90 Paul Street, London EC2A 4NE

British Library Cataloguing in Publication Information Available

Library of Congress Cataloging-in-Publication Data

Names: Epting, Shane Ray, author.
Title: The morality of urban mobility : technology and philosophy of the city / Shane Epting ; foreword by Lewis Gordon.
Description: Lanham : Rowman & Littlefield, [2021] | Series: Philosophy, technology and society | Includes bibliographical references and index.
Identifiers: LCCN 2021016508 (print) | LCCN 2021016509 (ebook) | ISBN 9781786608192 (cloth) | ISBN 9781786608215 (epub) Subjects: LCSH:
Urban transportation—Moral and ethical aspects. | Urban transportation—Philosophy.
Classification: LCC HE305 .E68 2021 (print) | LCC HE305 (ebook) | DDC 174/.93884—dc23
LC record available at https://lccn.loc.gov/2021016508
LC ebook record available at https://lccn.loc.gov/2021016509

Contents

Acknowledgments

Earlier versions of my ideas presented in this book that appear elsewhere have been revised and refined considerably in many instances. Perhaps they will continue to evolve. I suppose that is the way things go. Still, I am indebted to numerous people for engaging with me at conferences over the last several years, especially over coffee—where the real philosophy happens. These conversations helped me improve my thinking, gave me new perspectives in some cases, and guided my curiosities. Codirectors and executive committee members of the Philosophy of the City Research Group deserve special thanks: Michael Menser, Jules Simon, Ronald Sundstrom, Michael Nagenborg, Sanna Lehtinen, Judith Green, and David Woods. I appreciate the works of the philosophers, researchers, and brilliant thinkers whose works that I reference in this book.

I thank the journal *Environmental Ethics* for letting me use (and revise) my paper, "On Moral Prioritization in Environmental Ethics: Weak Anthropocentrism for the City," and the same goes for *Essays in Philosophy* for my article, "Transportation Planning for Automated Vehicles—Or Automated Vehicles for Transportation Planning?" (https://doi.org/10.7710/1526-0569.1635)

I thank Lewis Gordon for his foreword, wisdom, and kindness. Sven Ove Hansson and Diane Michelfelder deserve thanks for motivating me to write a book for this series. Andy Haight, Shelley Rios, Mark Bartos, Phillip Honenberger, Dave Beisecker, Carl Sachs, Remmon Barbaza, and Warren Stoltzfus—thanks. I appreciate the work of Frankie Mace and Scarlet Furness.

I dedicate this book to Ana, her family; our chihuahuas; and to the memories of my parents and my brother, J.K. Webb, and Coco Bravo.

Foreword

Shane Epting presents a fresh philosophical analysis of a problem worthy of study as the discipline continues to slug along through the twenty-first century with its gaze unfortunately locked in the past: How should we get where we need to go?

A discipline born from wonder about reality, much contemporary philosophy ignores its history of concern for the public interest as many practitioners imagine form over socially rich engagement with, proverbially, matters. What's the point of philosophy if it fails to address what it means to live meaningful lives?

This question of living meaningful lives inevitably leads to ethical and moral reflection. The scope of such efforts is, however, often suspect among at least professional philosophers. Should such concerns include reflections on power?

Despite a near allergy to the concept of power in ethics and moral philosophy, there is no hope for human beings without it. Lacking power, no one can do anything. This includes movement. Power, after all, means the ability to make things happen with access to the conditions of doing so. However ethical and moral one may wish to be, how can one be so without the ability and means of action—if even at levels of thought?

Although one can philosophize anywhere, the majority of philosophers prefer to do so in cities. Prime examples are Socrates bumping into fellow citizens for a philosophical conversation in the streets of ancient Athens through to Simone de Beauvoir, Frantz Fanon, and Jean-Paul Sartre sitting down for a marathon conversation in a café in twentieth-century Rome in 1961. In cities, philosophers are reminded of peculiarly *social* dimensions of philosophizing, despite Cartesian fantasies of self-absorbed acts of doubt, because of the fact that so many people live together in such places. It takes some time to reach one's neighbor in many urban environments, which would make philosophizing, in those cases, a lonesome and perhaps lonely endeavor. Cities are fertile grounds for philosophizing, at least socially.

Philosophy in the city is also a peculiarly political activity. It is, though not explicitly aimed at reflecting on power, one in which argumentation, disagreement, and commitments are governed by agreements to offer one's best without recourse to violence. In the city, for the sake of living together, speech is the primary means by which power takes the form of public empowerment; this activity is *politeia* or, as we say today, politics.

At the heart of sustaining political life is, then, the sustainability of cities, which here should be understood as places in which citizenship takes place. Although a city could spread across more than 100 kilometers or be as small as a single kilometer, the ability to reach one another swiftly shrinks the distance. The reduction of time over which to traverse space makes the largest of cities into tiny, intimate communities with possibilities of communication.

Transportation then functions as a concrete transcendental condition of citizenship and political life. Where transportation is marked by inequalities—including, for some, inaccessibility—the political consequences are fraught with injustices and, at the extreme, oppression. A society could extol democracy and freedom all it wishes; it would be meaningless for people who, smitten with aspirations of political participation, cannot get anywhere. In countries such as South Africa and the United States, this is evident in infrastructures designed for racial restrictions of white mobility and black immobility.

Technological developments are key for such movements. The wheel in antiquity reduced time and energy needed to reach from one point to another; in time, from ships to railways to airplanes, moving quicker across vast distances led to an ever-shrinking planet. Accompanying these were also information technologies ranging from the telegraph to wireless communication. Despite physical distances, these information technologies have ushered in a world of social access. Starkly evident during the COVID-19 pandemic, many people overcame physical distancing through being virtually close. The world continues to change, and in each instance, the unpredictable arises along with a slew of ethical and moral demands. Accompanying each technological development are new kinds of harms ranging from cyber bullying and theft to malfunctioning transportation systems with catastrophic consequences as witnessed in the airline industry from brand new passenger gets plummeting from the sky.

Understanding these challenges, Shane Epting assesses transportation as a right—since people's empowerment is premised on being to affect the world through reaching beyond their initial location—but he astutely makes the distinction between a right and a luxury. Although the movement and advancement of capital benefit from mechanisms that make movement expensive or stimulating transportation economies, the accompanying costs for planetary life demand pause and reflection. The imposition of old-style economies and moral ordering of life yield negative returns when brought to a global scale. A world of everyone expecting to live as kings and queens is, in the end, not a sustainable one. For the interest of ethical life—indeed, livable life—creative rethinking of technologies of movement, access, communication, and socialization is needed as, in addition to many other forms of life, many human beings are now struggling literally to breathe.

Drawing upon my theory of disciplinary decadence—which examines the problem of practitioners treating their disciplines as if they were complete and created by the gods—Epting points out that we are in error if we try to squeeze humanity into disciplinary and technological paradigms that no longer fit the ethical and moral demands of our age. Failing to assess the growing obsolescence—and dangers—of such models, the result would be an ignoring of reality in which hitherto viable models now produce suffering. To address this problem of cultivated suffering from epistemic stubbornness, Epting points to "intra-disciplinary decadence" in philosophy, wherein philosophers throw reality to the wayside in favor of the cult of subfield or professional membership. Philosophers doing this go against one of the greatest insights from antiquity—namely, Plato's *Allegory of the Cave*. Philosophers, after all, should be trying to get out of the cave instead of placing a boulder in front of it and urging everyone to walk back down into the delusions of shadows.

The French philosopher Alain Badiou elegantly pointed to an additional allegorical element in this famous example from Plato's *Republic*. He calls the activity of going back and forth to cultivate a community's exodus outside, simply, "politics."

Morality and politics are not, however, identical. While one could be individually moral, one can only be political through interacting with and affecting the world of others. A meeting of the two, however, is in the conduct manifested in political action. Where politics pertains to the expansion of power through the empowering of others, the inevitable path is to democracy. The proper norm for democracy, however, is one attuned to inclusion instead of exclusion. For Epting, this takes the form of "inclusive moral ordering." Made plain: people should have a say and be actively involved, with critical sensibilities, in the world that affects them, including building the foundations of the world to come. In place of the single-vision model of the transportation planner, Epting recommends processes of co-planning, guided by moral ordering. He repeatedly builds this argument through issuing a critique of various false dilemmas such as consequentialism versus absolutism or deontologism, of virtue versus societal needs.

The richness of Epting's analysis comes to the fore in the array of transportation problems and the technological innovations he analyzes. Can we, for example, live in societies of automated vehicles without changing who we are, the economies that result, and, as with all technologies, finding ourselves in the face of new problems to address? What are the forms of ethical attunement to foster in a world without forecast? Should not our commitments, our understanding of lives worth living, come into play?

As our planet continues to shrink, and the realities of how limited our resources are on this speck of dust we call the planet earth swirling through a cloud of dust that is our galaxy, the gravity of our situation is evident. It demands critical reflection—a hallmark of philosophy—to be responsible, and, beyond thinking in and about cities, taking on the urgent task of reimagining and changing them.

Such is the challenge posed by *The Morality of Urban Mobility: Technology and Philosophy of the City.*

Lewis R. Gordon

Chapter 1

The Road Ahead

This chapter's primary aim is to address the preconditions for the morality of urban mobility, a topic that is inherently positioned to cause trouble. I make this claim because transportation is a technology, and it helps to assume that the philosophical enterprise, in the form of well-ordered reason, is one as well. This notion holds we can think about it as a technology to help us address problems with transportation systems. To be clear, I am thinking about philosophy in this sense as a multifaceted conceptual process that helps us accomplish the task of thinking through the moral dimensions of urban mobility. In turn, while it might seem as if investigating the thought patterns behind transportation systems is not the best tool for the job, it is—if one seeks to fundamentally examine their range of statuses, from the ontological to the moral. Although this view cannot make the trains run on time, it can help us understand the significance of their operation and what it means for the urban condition. So, there is that.

The other goal is to provide an overview of the following chapters. Considering that the morality of urban mobility brings numerous sociopolitical elements into view, the motivation behind providing this outline is to establish a familiarity of the complex concepts that makes such an initiative manageable. The reality is that the morality of urban mobility has little to do with transporting people through the streets, and it has more to do with how we conceive of what it means to move in the city.

Bearing the latter point in mind, the primary takeaway from this book is not really about transportation. It is also not about the philosophical aspects that pertain to it as they intersect with peripheral issues about humankind. Yet, it does give special consideration to urban mobility due to the numerous moral dimensions that relate to its affairs. This focus's significance is that it encourages us to investigate areas associated with transportation systems the world over. It can provide suggestions for readers to think about mobility systems in ways that let them see their moral dimensions panoramically, bringing historical, social, political, and ecological elements into view. Engaging in this process could help us create cities that favor worthwhile goals such as socially just sustainability and human flourishing—which is the central message that I aim to convey.

One challenge in preparing this manuscript is bringing philosophical aspects into urban-mobility conversations in ways that do not betray established disciplinary traditions while not putting transportation specialists and enthusiasts to sleep. Another obstacle is resisting the temptation to explore philosophical "rabbit holes" as they are encountered throughout the following chapters. That is, as I touch on works in the history of philosophy, larger questions and debates lurk behind the scenes. To stay on track, however, those issues are best reserved for other investigations.

In turn, this book spotlights numerous concerns that connect to transportation, but they may not be immediately known to all or most people. While examining mobility in this particular manner might appear to discount elements of paramount importance for planning professionals, my aim is not to challenge their authority.[1] Instead, it is to illustrate that any such decision has several ethical concerns that only occupy a space on the margins of our minds for any number of reasons. One task is to bring visibility to the topic, revealing the importance of caring about urban mobility beyond being a passive participant. Still, it is not in everyone's interest for a full-fledged urban army of amateur planners to control the streets. The goal here is to discover an adequate degree of moral allurement that can balance the former and the latter.

Due to this reality, there are no direct claims about how to "do" transportation engineering or planning beyond the philosophical enterprise. That job is for professionals who can keep us safe from the dangers inherent to transportation systems. In turn, this book is not a case of an outsider looking in, telling professionals how to do their jobs. Yet, there is no good reason why anyone should not give extensive commentary on urban mobility, considering that such undertakings increase conversations about it. Transportation systems affect all urban dwellers and travelers, and our experiences count.

The above claims entail that there are no suggestions that endorse any particular mode of transport per se. Instead, the case presented in the following pages champions mobility systems that promote worthwhile aims such as those mentioned above, which are, at the same time, areas of discussion subject to debate. The goal of undertaking this project is not to confront professionals or people who favor particular transport modes. Instead, it is to invite them to spend some time devoted to examining the moral enmeshment that is transportation in the city, assuming that they are not aware of it. It also invites philosophical purists searching for content that applies lessons from the discipline to problems in and with the streets. Philosophers participating in the expanding research area of "philosophy of the city" have demonstrated that topics such as urban aesthetics, urban technology, and participatory democracy in municipal affairs are connected issues. This work intersects with many such subjects. Aside from these groups, I hope to entice people to use this book to gain perspective on what their city's transportation system means for them and their ways of life. They are the ones who can harness urban power to transform the cityscape.

Another aim is to reveal how transportation systems remain entrenched in a global web of fluctuating conditions that help shape and reshape the outcomes that we can find associated with urban mobility—once we have an extensive view of such situations. Engaging in this exercise reveals the character of the myriad, intertangled ethical concerns that come from the need to navigate urban landscapes. Due to such a reality, one result of fleshing out this project is to show that the entirety of transportation concerns, as they remain situated in broader socio-material contexts, have numerous tensions.

This situation makes it incredibly challenging to create mobility networks that are free from critical problems. If we accept this notion in its fullest form, then we need to accept that transportation justice is always in a "toward" mode, as if it were a mirage in the distance that moves further away as we approach it. This position brings an inherited view of justice from the philosophical canon into question, one that sees justice as somewhat of a fluid concept. This notion suggests that we cannot attain it (probably). However, the task of working for it is intrinsically good and instrumentally valuable in the sense that it can provide relief for people who are suffering or who could fall victim to the harmful conditions associated with transportation systems. By default, the goal is progress, not perfection.

This outlook requires shifting our focus away from rigid solutions, such as the typical moral framework that exists most fully in a perfect world, toward one that correlates to reality. The irony here is that the vast majority of research in this project remains theoretical. It holds steady as an enterprise that is widely applicable to cities across the globe. These two notions seem to be at odds. They are. They create a tension that I employ to address the complex nature of transportation issues throughout the text.

On the one hand, although there are several ways to approach these troubles with established frameworks, they all suffer from a shortcoming that, through their particular branding, cannot yield transportation justice that remains free from substantial objections. Rather than strictly relying on established philosophical positions, gaining a clearer perspective of the encompassing nature of transportation affairs might spare us from repeating some of the past problems. That is, appealing to a conventional approach or proposing yet another framework in the traditional sense is unlikely to deliver a pathway to transportation justice.

This point does not suggest that this book entirely dismisses the *utility* of frameworks. Still, it does ask us to approach them suggestively rather than looking at them as sacred tools that we must employ within a set of defined, philosophically pure parameters. Yet, I am not advocating for this view for the mere sake of rebellion.[2] Instead, I favor it because it is suitable for dealing with real-world transportation issues that require flexibility. Many scenarios that exist or will emerge would benefit from insights from the frameworks in moral theory's history. Yet, the ways that we need to apply them will upset strict adherents to canonical orthodoxy.

Considered in such a sense, one could call the proposal that follows an "anti-framework" framework, suggesting that the outlook that I am putting forth is inherently inconsistent. I am fine with that label because it fits with the world, which is often how things work out. At least I am consistent with my affinity for inconsistency. Despite any such incongruence, I still do not want people to suffer from transportation ills while we figure out the cosmological structure of the universe and how it equitably relates to scheduling bus services.

The underlying worry that comes with accepting and embracing the position I advocate is that it smuggles in sympathies for moral relativism that deny moral absolutes. In turn, we have to stack the possibility of delivering short-term solutions against the chance that it could lead to atrocities. If it turns out that searching for mitigatory efforts directly leads to any such violence, then we will have much more pressing issues than the affairs of urban mobility.

Aside from those topics, the practical reality that accompanies my position suggests that there is no definitive standard that we can attain, but we must continue to strive for it. Although pushing the mobility-justice boulder up the hill holds steady as an indefinite task, the scope of possible impacts demands it. Some such affairs are well known, while other areas are stealthy. Still, they affect life in so many ways that call for attention from people with backgrounds from across the academy, philosophers included.

For instance, these matters impact numerous categories of stakeholders. Many of these issues are ingrained into our environs' social structures, and we may not be aware that they exist—or that there are alternatives we can embrace to deliver better realities. These topics have a wide range, from fuel production and the global market, all the way to the urban dwellers who depend on buses, rails, scooters, and sidewalks to navigate the cityscape. With so many considerations to weigh, one could say that they form a "moral mobility enmeshment." It is a reality that deserves attention, even though it demands thinking in a way that requires us to blend the abstract with the concrete.

Although we cannot always see the connections between these two spheres, they are ones that numerous people know through how they impact their daily experiences, which, over several decades, shape their lives. These matters include people spending time with their loved ones, getting a primary education or advanced degree, or simply relaxing after a difficult day at work. If we expect people to pull themselves up by their bootstraps, we need to ensure that they have a way to get to where they are supposed to go, once they are wearing their boots, assuming that they can afford the footwear (or have practical access to it).

Aside from this point, due to the manner wherein mobility blends into the backdrop of people's everyday routine, these facets of life go largely unexamined as concerns for moral inquiry, even though they may be the source of other forms of investigation. In turn, such affairs do not appear as ethically troublesome conditions to many people, meaning that it could seem to them that there is no need to improve mobility systems. Yet, once we start looking at transportation through a moral lens, numerous problems come into view. On the one hand, we have to deal with some of these problems as they emerge, from city streets to suburban cul-de-sacs. On the other hand, the attention that we give to each issue says a lot about prioritizing our values—and the quality of humans' lives.

For example, although one might suspect that housing, food security, or lack of education are the only reasons people who are stuck in poverty cannot escape it, transportation is a prime offender.[34] It is entirely unrealistic to expect people to self-determine the means for social and economic advancement if they are spending, say, three to five hours per day going from their home to their job or multiple part-time jobs. This notion indicates that vulnerable people, marginalized groups, economically disadvantaged communities, and seniors on fixed incomes will continue to suffer because of transportation systems that were not designed keeping them in mind—or at least beyond codified protocols that establish basic thresholds.

Thinking about this notion in a global context means that billions of lives are determined by where people can afford to live and the distance and transport modes they can use to get to work and back to their homes. One can fathom that many of these people have had to make tough choices between essential purchases such as medicine and fuel due to mobility concerns. Along with these decisions that affect their lives, the "where" element of the places they call home at times gets caught in transportation planning's crosshairs. This sad reality magnifies the extent of mobility problems. That is, neighborhoods belonging to marginalized communities have an unfortunate history of being destroyed for new highways and roadway expansion projects.[5]

Along with such considerations, the after-effects of building expansive roadways could continue to worsen circumstances, extending their durations indefinitely. As mentioned above, this situation normalizes the horrid conditions of inner-city driving that become accepted, with or without resistance. For instance, due to highways' widespread nature, people in numerous countries are forced to "fight" traffic every day, which is not an easy battle. The arrangement of these mobility modes also creates conditions that manifest unique ways that affect human behavior, such as road rage. This phenomenon is not uncommon, and research shows that up to one-third of all drivers engage in it at some point in their lives.[6] Such outcomes would arguably not exist, or perhaps incidents would be far less common if our roadways' conditions were improved to combat such effects.

Aside from giving attention to human lives, transportation also plays a role in climate change, impacting human and nonhuman life on the planet.[7] When considering the impacts of urban expansion, wildlife becomes displaced for our mobility as an indirect consequence.[8] They wander into our cities. We treat them as trespassers. Along similar lines, over a million (nonhuman) animals die each year in many countries on our roads, becoming roadkill.[9] Due to this actuality, road ecology is now an area of practice, training, and study. Adjustments are made to reduce road-related deaths of wildlife, but one could argue that the motivation behind such decisions is to preserve human life and property, a notion that I explore in great detail in the following chapters. Yet, along with the biotic community, the abiotic landscape also succumbs to our needs to move about our cities and the planet, forsaking the wisdom from Aldo Leopold. He once taught us the lesson that would become the backbone of today's environmental movement: "A thing is right when it tends to preserve the integrity, beauty, and stability of the biotic community. It is wrong when it tends otherwise."[10]

Considering that paying attention to nonhuman life and resources will affect generations of people who do not yet exist, our transportation choices will affect them in the distant future. This notion suggests that we cannot simply make transportation decisions based on humankind's immediate mobility needs, even though this issue is riddled with theoretical challenges (which unfold later in the text). Still, we must also think further ahead, perhaps looking to several generations of people who will have to contend with the decisions made decades ago. Suppose one considers that today's people wish that they did not have to live with global climate change, harmful urban forms, and respiratory illnesses associated with transportation. In that case, the argument that we do not wish to harm people who *should* exist in the future should not be such a hard sell.

Although these considerations deserve attention when addressing transportation affairs, neighborhoods, historic architecture, public parks, common spaces, views of waterways, and commercial and industrial districts are also instrumentally and intrinsically valuable. These conditions require that we should bring them into perspective to see how decisions that pertain to transportation will affect them, too. Yet, preserving these artifacts comes at a price, which means building new transportation infrastructure and supports will require advanced study to see how much value they have when stacked against immediate and future mobility needs.

All of the areas of concern mentioned above, while not necessarily exhaustive, illustrate that the picture of designing a transportation system that is inherently just is complicated in the best case, and impossible in the worst case. Such is the nature of the mobility beast. Complex human environments that mostly remain in flux make it necessary to have transportation systems that help create the conditions that we must always maintain to produce the outcomes that arise from these situations. Despite the complexity that we find with such scenarios, mobility modes are much more than ways to move about the city. Due to these circumstances, they require the kind of attention that can account for such conditions.

For example, in some cases, they are markers of identity. They bear cultural, ideological, and individual significance. People are loyal to brands of automobiles, such as Subaru and Toyota.[11] They are also dedicated to particular personal transit styles such as lowriders and pick-up trucks, along with other branded-loyal modes of transport such as Harley Davidson motorcycles.[12] While private vehicles have followers who strongly identify with such machines, different modes of transport also have devotees.

Consider, for instance, that there are people who depend on bus services who view ridership as belonging to a secondary mobile community.[13] Other users could have political identities associated with ecological considerations due to climate change's moral aspects, favoring environmentally minded models such as hybrids and electric cars. Along similar lines of thought, groups exist that strongly prefer bicycles. For instance, Portland, Oregon, has pushed against the transportation status quo, opting for advanced infrastructures so that bicycle riders can travel on bikes with more ease (despite unfavorable weather conditions).[14] There are incredibly deep-rooted commitments to bicycle ridership, and now there are apartment complexes without parking spaces for vehicles but plenty of room for bicycles.[15] While creating bicycle-favoring living spaces and communities has immediate benefits for people who are partial toward life on two wheels, such a trend also pushes against established traditions in a way that goes beyond immediate mobility. It shows how focusing on one dimension of urban mobility can have far-reaching impacts on other areas of life, such as housing. This point indicates how changing and encouraging transportation dynamics can bring different aspects of urban living into view that can positively impact cities. This notion is why this book focuses on transportation.

Yet, attempts to change transportation systems have to confront the realities that many users of various stripes have strong feelings toward their personal choice of transit, even when such cases do not lean toward the benefits mentioned above. These conditions could complicate matters to alter existing transportation systems, which industry leaders or transport authorities could deem as avenues to progress. These professionals' approach must bring such preferences into consideration to achieve acceptance, which is necessary for success. If the measures that they adopt remain inclusive to the point that mobility loyalists are willing to change because they are given better options, resistance could transform into enthusiasm.

Consider, for example, automated vehicles. Suppose brand and mode loyalists see them as far better technologies than their current preferences. In that case, we could see people abandoning their personal transportation choices faster than they gave up their privacy for access to social media. This view also indicates the perpetually changing character of the urban sphere. Cities are always changing. Keeping pace with the matters that emerge from these scenarios will not end. For example, although we must address issues that already exist in our cities, we must also deal with problems as they crop up, along with the issues that we can anticipate.

These conditions demand that we are attentive to the changing needs of societies that depend on mobility networks, meaning that we must engage with them on a case-by-case basis, as each instance affects entire transportation systems. This point suggests if we can consider that transportation issues occur across the world, and we can deal with them on a smaller scale, a few more steps in this sequence would indicate that there is a pattern at play. We can identify and work with it to produce desirable outcomes. Such a notion means that we need to look at the parts of each instance to determine if they are germane to a mobility problem. Employing this approach, addressing concerns that break issues down into part-to-part and part-to-whole relations, provides a way to pinpoint particular uniquely embedded problems in distinct metropolitan centers.

Gaining an efficient, albeit effective manner to deal with mobility concerns will provide a platform for launching efforts to address transportation ethics, as mentioned above, despite not all issues being equal. This latter notion is paramount because while there are numerous ethical issues, some matters should concern us more than other topics. In turn, developing a way to deal with "moral ordering" is the product that the thinking behind this book aims to achieve, along with the outcomes that accompany it. Clarifying these points is the task ahead. To put it briefly, for now, it holds that while transportation systems affect several groups of stakeholders that deserve moral consideration, the ways that we deal with them speak volumes about our priorities, even though defining "our" is a separate matter. The reality here is that some groups must be prioritized beneath other categories, and such practices require a fairly nuanced explanation, especially when the reason "why" is not immediately obvious.

With an understanding of moral ordering and its significance for pursuing worthwhile goals such as those mentioned above, the message here is that through gaining an advanced understanding of the ethical dimensions that remain connected to transportation systems, (some) readers of this book will be challenged to think about mobility systems as a means to improve the quality of human life. To reiterate an earlier point, gaining this perspective does not require a new framework as much as it benefits from a shift in attitude. Why? Though one could argue that employing a new framework could decrease the chances of committing unethical acts or worsening existing problems, a framework—no matter how rigorous its proponents claim that it is—is only as strong as its shortcomings. These are the times when it "fails." Using one could increase reliability and predictability, which can identify and understand the complexity behind decisions that will impact billions of people, the environment, future humans, and elements of the cityscape that people hold dear.

Strict or "strong" moral frameworks are too ambitious to deal with such serious problems. The difficulty of sticking to a strong framework is that situations can arise when we want to go against it. If we employ a strong framework, doing so carries the idea that it is "strong" because we cannot stray from its instructive orientation. This criticism is common for strict moral theories such as deontology in its original conception. Although it can deliver guidance time and again, there comes a time when it backfires, sticking us with unpalatable outcomes. An alternative is that we go with a weaker framework that still provides guidance, but it has an inherent flexibility that allows it to bend to situations that call for it. Weakness, in this sense, becomes a strength.

One could push back against this claim, asking: What is the point of employing a "weak" framework if its underlying purpose is to deliver "strong" directives for action, ones that will remain unwavering in their commitment to the tenets of the framework? This question implies that adhering to a moral framework in a real-life situation such as when making a transportation decision could become a source of oppression when adhering to the framework goes too far, giving us outcomes that we do not desire. Rather than proposing a framework in a strict sense of the term, this book aims to inspire attitudes for moral ordering, paying attention to the list that I previously mentioned, in the *suggestive* order that they appear here: vulnerable and marginalized populations, the public, nonhuman life, future humans, and anthropogenic urban artifacts.

I use the term "suggestive" above to underscore (again) the idea that we are not dealing with a strict framework, which is a primary motivation to avoid the norm of developing and then employing frameworks for matters that begin with theory and then transition to practice. That is, there is an underlying tension between the philosophical proclivity toward striving to deliver absolute moral principles, as mentioned above, that are often thought to apply to any given situation (e.g., trolley problems). If this is the case, then academic philosophy remains incongruent with the real world where people suffer from respiratory illnesses because we cannot change mobility systems quickly. This notion only indicates that there are much larger systemic concerns wherein numerous transportation issues blend into the urban sphere's backdrop.

For instance, how can we defend the claim that people who exist now, who are suffering due to incredibly long times spent in transit, should continue to suffer because they are doing so for the sake of people who do not exist yet? I am not saying that such a claim is indefensible. Rather, I am pointing out that holding this position requires thinking beyond a cursory glance about the need to align the defense of a decision or action with a framework, one that does not have to yield to the restrictive nature of a framework. Lewis Gordon might refer to this situation as "disciplinary decadence," considering that some philosophers could want the world to conform to their ideological prescriptions for it instead of adjusting their beliefs to reality.

Here is where we can draw from Gordon:

> *Disciplinary decadence* is the ontologizing or reification of a discipline. In such an attitude, we treat our discipline as though it was never born and has always existed and will never change or, in some cases, die. More than immortal, it is eternal. Yet as something that came into being, it lives, in such an attitude, as a monstrosity, as an instance of a human creation that can never die. Such a perspective brings with it a special fallacy. Its assertion as absolute eventually leads to no room for other disciplinary perspectives, the result of which is the rejection of them for not being one's own. Thus, if one's discipline has foreclosed the question of its scope, all that is left for it is a form of "applied" work. Such work militates against thinking.[16]

We can employ Gordon's wisdom to identify the kind of thinking that turns against the enterprise of untethered yet carefully attended exploration. Applying this insight to philosophy itself, wherein philosophers attack others in the discipline for their approaches to investigation, I enjoy calling it "intra-disciplinary decadence." On this note, to argue that we *must* employ such a framework to count as "philosophy" is dogma, which allegedly goes against a sound philosophical enterprise's nature. While numerous frameworks exist, the following chapters explain why they need replacing to deal with highly specialized affairs, especially considering that the quality of people's lives is at stake and the numerous matters that stem from mobility systems. Taking this point seriously means that philosophically investigating such issues requires positioning oneself against the decadent perch, pushing against views holding that philosophical thinking always provides a clearer picture of reality because the discipline grounds other areas of study. In turn, attempts to discuss the conditions for transportation justice and ethics must adhere to the participatory structure of restorative measures that are inherently inclusive.

This point entails that people who have been harmed by the socio-material arrangements of transportation systems must inform moral ordering. Without it, any effort is wishful thinking—even if the results still work out because one of the consequences is that people who have been harmed still lack the ability to participate in such decisions. Bearing this point in mind, there are larger issues at play that highlight how mobility specialists' work has dimensions that we do not ordinarily consider. At the center of these affairs is that there is a battle for the future of urban mobility. While there are numerous urban planning conceptual "isms" that seek to repave the urban sphere with more just or equitable means, some industry leaders want to provide a future for mobility that puts their profits at the forefront. Such a move would consequently place all other options as secondary choices, which must work around these future technologies.

Of course, I am talking about automated vehicles and their role in urban mobility if they ever materialize as numerous predictions promise and re-promise. Although automated vehicles have only made minimal appearances in our cities thus far, the bark of their proponents is so incredibly loud that we cannot ignore the possibility that they might one day bite. Despite their lack of wide-scale deployment and their perpetual postponement, the anticipation behind their future success has already made a presence in the transportation-planning literature.

Critiquing this situation on the meta-level, the thrust of Karel Marten's primary criticism that transportation plans often favor the future over the past mistakes continues to ring true.[17] This notion indicates that we are not merely dealing with one kind of real-world concern, but at the same time, we must also confront how we conceptualize the patterns behind the thinking on urban mobility. The problem is not just that the trains do not run on time (in many instances), but such a case is merely emblematic of larger ills that govern our thinking on such matters. Pursuing these technologies with unyielding enthusiasm while taking nary an interest in long-lasting transportation injustices might be the way that the world works. However, that notion says nothing about how it should be, and it does little for illuminating alternatives that could deliver better outcomes.

Although the reasons above paint with incredibly broad strokes, they illustrate the scope of the work required to secure a more socially just future for urban mobility. For the planners and engineers on the streets who are motivated to address these concerns, each new bike lane that improves safety and every sunshade added to bus stops to protect riders from the sun's unforgiving rays help each of them. Many of them are heroes.[18] When taking on mobility problems in real time, they must (knowingly or unknowingly) encounter the entirety of urban mobility when they approach an individual instance.

Although it might appear as if they are addressing an elemental concern such as bicycle safety, they glimpse the totality of their transportation system. If these professionals can improve the outcomes associated with such a seemingly insignificant problem, they can make incremental progress toward improving urban mobility. Bearing in mind that individual transportation affairs are part of the global matter of climate change and its associated ill-effects compound, many such persons and groups are working with the odds against them. One could argue that developing solutions to anticipated challenges while mitigating harms that fall outside an assignment, or dealing with an immediate mobility issue that bolsters efforts to mitigate climate change, could qualify as supererogatory actions. For some such tasks, transportation professionals could deserve moral praise.[19]

Consider, for instance, that their commitment to such efforts puts them in a position that makes them extraordinary. While such a practice is unusual, this circumstance says more about social norms than praiseworthy practices. Why should it not make sense to spotlight and appreciate transportation specialists who work to improve people's lives? Why should calling them heroes come with hesitation—due to anticipated ridicule for employing such a term in this fashion?

From my perspective, these questions are rhetorical, but, as mentioned earlier, transportation systems blend in with the cityscape, meaning that they usually go unnoticed. If these technologies are viewed without much regard, then the people hired to make them operate might also go without accolades for "just doing their jobs." This point is fair. However, considering the sheer number of lives and livelihoods at stake when thinking about the collapse of such systems, one could argue that these occupations take a special kind of courage, one that goes mostly unacknowledged. People do not cheer when the trains run on time, but they tweet about it when they arrive late. Bus drivers in some cities receive thanks from riders daily, but I would be surprised by transport engineers receiving "thank-you" cards. Yet, touching on these aspects merely begins to uncover the intricate nature of their professions.

Still, deserving praise for this sort of work goes beyond mere gratitude for good services rendered. To fully appreciate that there are grounds for supererogatory praise, we need to comprehend the complexity of their work when taking on problems that involve issues from the past and present, and those that will have effects long after they live. Understanding this dimension of their work requires extensive unpacking.

For instance, they must balance multiple ethical considerations. Essentially, having several categories of stakeholders complicates decisions about whom or what should receive consideration, what degree they should receive it, and how to deal with competing interests between the kinds of entities that deserve moral consideration. To address these affairs ethically, I suggest moral ordering as a conceptual device to help them deal with transportation issues that involve multiple stakeholders. The attempt here is to think through these issues, providing a perspective that focuses on the ethical situations that arise from trying to deal with the problems above, revealing why we should look at transportation issues as one of the most pressing moral issues in society. After making this case in the following chapters, I hope that mobility gains respect as a topic of interest in applied ethics, replacing trolley problems with the problem that there are no bus rapid transit lanes in places that need them.

Along these lines, the first aspect that requires acknowledgment is that we are ultimately concerned with outcomes, making this anti-framework inherently consequentialist. We are dealing with people's lives. Trying to stick to one's duties or develop moral character are fantastic goals, but I would rather orient attitudes toward mitigating harm as a priority. This point should remain unapologetically undeniable. Yet, the only means that exist for dealing with such matters are frameworks. Does this condition mean that I have already shown that the task I have discussed thus far is already doomed? No. However, it does suggest that there is something significant about frameworks that we need to rescue—and there are some things about frameworks that we should discard.

In terms of the former, the idea is that with each framework (i.e., moral theory), they reveal aspects of the situations that challenge us, and transportation is no exception. Holding on to these ideas, saying that we should promote people's happiness or respect individual rights, is vital. We should keep these ideas firmly in view, but the tendency to codify them or view them as anything more than suggestions will backfire *eventually*. If we consider these two points in tandem, then a view emerges, showing that we can employ moral theories in ways that make good use of their insights while at the same time not taking them as a dogmatic, philosophical gospel.

The problem with this view, as mentioned earlier, is that it moves toward moral relativism. One could argue that discussing morality without appealing to absolutes is futile for some people, collapsing into subjectivism. This concern appears serious because it could lead to typical objections that consequentialism could justify immoral means to good ends. However, it is doubtful that transportation professionals will engage in the kind of behavior that would qualify as war crimes unless you consider that building massive, global mobility systems that play a significant role in changing a planet's climate fits the bill.

That notion aside, the problem with that objection is that it follows the same lines of philosophical reasoning that created strict frameworks in the first place. For instance, though there will always be the possibility that a municipality could fuel its transport system by ill means or some such scenario, do we need to dedicate time to such thinking while there are real-life issues that require solutions? No. With this point in mind, we can tinker with frameworks, removing parts and pieces to advance our thinking, but we must do so in a way that recognizes the inherent limitations of those frameworks, looking at them as conceptual tools to help us get the job done. As indicated above, the goal here is to remove the clutter, the parts of frameworks that hinder our thinking.

In this sense, analyzing the difference between a regular motorcycle and a chopper can be useful.[20] For instance, a standard motorcycle has been designed and built for the typical motorcycle rider. It has numerous parts that help it run smoothly and efficiently. It could belong to anyone. Yet, a chopper is different. Several parts have been removed, "chopped." It can now go *faster*. Instead of being a one-size-fits-all vehicle, keeping the essential parts and adding customization suits the individual who will ride it. In this case, the task is to develop attitudes conducive to building and/or rebuilding *specific* transportation systems that improve the urban conditions that influence the qualities of diverse human lives. Due to this reality, we need to chop several frameworks, saving their central ideas and discarding any notions that they are absolute in any shape, form, or fashion.

Considering that transportation systems are technologies, the specific kind of consequentialist approach behind moral ordering is structural ethics, which comes from the philosophy-of-technology literature. It holds that we need to understand technologies as playing "moral roles" in more extensive socio-material arrangements.[21] We can say that a technology plays either a good or a bad role in relation to the outcomes that unfold when the technology is involved.[22] As this story plays out, I favor thinking about individual technologies as parts in terms of their relations with other transportation parts, along with the whole transportation system, one wherein there is no need for discussions about nonhuman agency. This view pushes away from popular positions in the philosophy of technology, which hold that technologies have some kind of agency.[23] Instead, the focus on moral responsibility rests entirely on humans. Although this text employs works from the philosophy-of-technology literature, it chops views of nonhuman agency because they do not do anything to advance our thinking in relevant regards. They only slow us down.

While this approach provides a way to think about outcomes that transportation systems help produce, it also needs to consider the other categories mentioned above. This notion suggests that we have to deal with competing interests between the categories when designing, planning, reconfiguring, and maintaining mobility networks. Although it would be idealistic to employ universal design principles or a similar approach to plan and troubleshoot, transportation issues do not solely exist in ideal situations. This idea implies that we need a reliable yet flexible means, providing a way to make tough calls that entail that one or more categories receive preference—along with a straightforward way to justify the thinking behind such decisions.

For competing interests, one of the most problematic areas concerns the interests between humankind and nonhuman nature. Considering that these affairs ultimately deal with outcomes that affect humankind, a position developed in environmental ethics known as "weak anthropocentrism" is inherently consistent with structural ethics. Put briefly, this position holds that humanity is the prime ethical consideration for decisions, but we can also make concessions for the nonhuman world without only appealing to instrumental values.[24] That is, we can include nonhumans into our moral prerogatives for their own sake, their intrinsic values. Bearing in mind that structural ethics is concerned with outcomes, including weak anthropocentrism means that we have to include nonhuman nature into the scope of moral considerability.

Yet, weak anthropocentrism does not make any claims about who or what deserves any consideration regarding prioritization. In turn, when it comes to the first category, vulnerable people, conceived of broadly, should be the primary focus for action. The idea here is that they require an advanced solution that acknowledges their situations, which means that efforts to support them should be prioritized when developing transportation measures for the public. If we can create incredibly advanced mobility systems, it only seems reasonable to not cause or perpetuate harm or injustice. On the contrary, such systems should count as instances to make their lives easier, to facilitate human flourishing.

When it comes to the general public, decisions concerning transportation should balance systems that appeal to people as individuals and society's betterment as a collective whole. To balance these considerations, the central idea here is that we want to provide people with the transportation choice that fits their lifestyle, but we don't want it to impede the collective ability of people to travel without being inconvenienced to the point that it causes harm. Moreover, we also do not want the result of this balance to blowback on vulnerable or marginalized people.

Depending on the circumstance, our dealing with mobility matters should also not cause incredible environmental destruction, either immediately or over a long duration. This notion implies that we want the nonhuman environment to be there for future generations. We hope that these individuals will exist, meaning that we must meet the conditions for their eventual presence when dealing with an issue such as transportation so that we do not reduce their chances of living. Despite this motivation, it is a challenge to plan for them when there are suffering people. In turn, arguing on their nonexistent behalf remains challenging, but we cannot outright dismiss such concerns.

Lastly, after acting (i.e., planning, building) in a manner that prioritizes the above categories in the order that they are listed in most instances, considerations should be made for neighborhoods, architecture, and other urban artifacts that residents or other groups hold dear or depend on for a variety of reasons. This notion does not entail that every such artifact requires preservation without reservation. Still, it does hold that, similar to works of art or any other intrinsic-value bearing entity, they deserve a place in the taxonomy described above. Yet, bearing in mind that most of these things are replaceable, meaning that they are not entirely essential in a high number of instances, one can make a case showing that they do deserve consideration, but when stacked against categories such as human life, we see that their place is lower on the scale.

Although one of the primary benefits of the classification above is to show that we can provide a theoretical device (moral ordering) that can be of service when making decisions that pertain to transportation, it is not absolute. While it gives an order of guidance, that, when not followed in many instances, it becomes a problem of moral prioritization. We want to claim that we have obligations toward some people over any alleged commitment to another entity, such as rats. Yet, there could easily be a case wherein we should act for nonhumans, seemingly acting for them over us. This notion means that it is a weak system of moral ordering, which holds that any ordering is only moral under a particular set of conditions, even though we must appeal to broader moral theories to bolster our decisions that pertain to highly individualistic choices.

This complex dynamic is another reason I resist the urge to position moral ordering as a traditional framework. While one could argue that being a weak system diminishes its capacity to provide predictability, which would reduce its worth as a conceptual tool, I counter this claim, holding that this quality makes it flexible. This feature makes it adapt to unique circumstances, and each city and transportation system has its own identity. In turn, having a theoretical device that can shift to accommodate new surroundings is a fitting attribute, not a liability. This characteristic's primary benefit is that it allows for lower-ordered entities that typically receive less consideration to receive additional attention and respect when required. The same idea applies to high-ordered entities, which will not always warrant prioritization.

As mentioned earlier, this characteristic, this weakness, is the strength of this approach, this anti-framework.

The challenging aspect here is identifying the conditions that require us to violate the order of moral prioritization. That idea is addressed throughout this book, paying attention to a few select works in the history of philosophy that allow us to follow this line of thinking trajectory, applying it to transportation as it appears in cities across the world. In that sense, moral ordering and the problem of moral prioritization are enduring conditions with a long shelf-life. Although such issues remain highly applied, this aspect exhibits that we must balance real-world concerns with the theoretical dimensions that fundamentally underpin them.

In the chapters that lie ahead, I flesh out this view. It begins by examining some of the harmful conditions that have emerged in the transportation sphere, along with the problems of employing frameworks. To ground this view in a way that provides us with the ability to examine transportation systems at their most fundamental level, as individual parts that are also parts of a larger whole, I borrow and take inspiration from mereology, a field of study that explicitly deals with parthood. Next, I turn to moral ordering as a way out of these situations, walking through each consideration. These chapters position us to see the possible advantages that this process provides. Such benefits can transform the methods that shape transportation systems. If carried out effectively, such measures hold the potential to fundamentally restructure urban mobility to support worthwhile goals such as socially just urban sustainability.

As mentioned earlier, one of the most active areas of multi-disciplinary and interdisciplinary research into the future of urban mobility focuses intensely on automated vehicles. Although the reality of seeing these vehicles dominate the cityscape is not on the immediate horizon, examining such possibilities and eventual actualities provide an opportunity to explore some of the many considerations that must factor into how we implement them into our urban environments. Along with this point, municipalities and transportation authorities must also thoroughly examine the inquiry: Who gets to weigh in on such decisions?

The fascinating notion behind this idea is that meaningful participation in such affairs is seen as an element that could complicate or harm efforts to improve mobility in our cities. While this point deserves significant time and attention, it also reveals one of the future avenues for transportation specialists to pursue to deliver outcomes that favor worthwhile goals such as socially just sustainability. In turn, we should welcome such undertakings with gusto.

Along with this point, I examine some of the possible avenues that mobility could take, wherein I advocate not for a particular mode of mobility but push for the helpful attitude that will allow us to keep working toward transportation justice. This idea rests on many of the areas of study that contemporary mobility thinkers advocate for, aiming to diversify transportation worldwide. If we genuinely desire mobility networks that do not resemble the past, we must look beyond the thinking patterns that limit our outcomes.

Although we need to create the conditions that will allow us to move beyond our current transportation systems, the more pressing need is to transcend how we conceptualize mobility. Hopefully, the work here counts as steps in that direction. Considering that cities will always require mobility, the road ahead will continue indefinitely. Despite the reality that this view suggests, it also implies that thinking through our mobility problems requires several lenses to see their totality, along with what transportation systems have the potential to become.

NOTES

1. Throughout the text, I will use terms such as "transit," "transport," "transportation," and "mobility" somewhat loosely and interchangeably because I am not using them in a technical capacity in most instances.

2. The pattern behind this kind of thinking is along the lines of thought in business ethics. For an example, see Al Gini and Alexei Marcoux, *The ethics of business: A concise introduction* (Lanham, MD: Rowman & Littlefield Publishers, 2011).

3. Evelyn Blumenberg *et al.*, "Getting around when you're just getting by: Transportation survival strategies of the poor," *Journal of Poverty* 18, no. 4 (2014): 356. This article provides an expansive literature review that investigates the complex nature of how transportation issues present numerous challenges for people who are already financially burdened.

4. Martin Wachs, "Transportation policy, poverty, and sustainability: History and future," *Transportation Research Record* 2163, no. 1 (2010): 5.

5. The history of the Bonton and Ideal neighborhoods in Dallas, Texas, serve as examples. For more information, see Briana Payne, "Oral history of Bonton and ideal neighborhoods in Dallas, Texas," M.A. Thesis, University of North Texas, 2015.

6. Randy Sansone *et al.*, "Road rage: What's driving it?" *Psychiatry* 7, no. 7 (2010): 14–15.

7. Soo Chen Kwan *et al.*, "A review on co-benefits of mass public transportation in climate change mitigation," *Sustainable Cities and Society* 22 (2016): 11.

8. Chun-Wei Huang *et al.*, "The importance of land governance for biodiversity conservation in an era of global urban expansion," *Landscape and Urban Planning* 173 (2018): 44.

9. Andreas Seiler *et al.*, "Mortality in wildlife due to transportation," in *The ecology of transportation: Managing mobility for the environment*, eds. John Davenport and Julia Davenport (Dordrecht, NL: Springer, 2006).

10. Aldo Leopold, *A Sand County Almanac and sketches here and there* (New York: Oxford University Press, 1949), 224–5.

11. "Brand loyalty increasing among new-vehicle buyers, J.D. Power Finds," *J.D. Power Press Release*, July 15, 2020. https://www.jdpower.com/sites/default/files/file/202007/2020046%20U.S.%20Automotive%20Brand%20Loyalty.pdf.

12. Reid Heffner *et al.*, *A primer on automobile semiotics* (Davis, CA: Institute of Transportation Studies, University of California, 2006). https://escholarship.org/uc/item/3577d04p.

13. Jeff Nash, "Bus riding: Community on wheels," *Urban Life* 4, no. 1 (1975): 99–100.

14. Peter Pelzer, "Bicycling as a way of life: A comparative case study of bicycle culture in Portland, OR and Amsterdam," in *7th Cycling and Society Symposium*, pp. 1–13 (2010): 4.

15. Stephen Beaven, "New Portland apartment buildings with no parking have neighbors worried about congested streets," *The Oregonian*, July 12, 2012.

16. Lewis Gordon, *Disciplinary decadence: Living thought in trying times* (New York: Routledge, 2015), 5.

17. Karel Marten, *Transport justice: Designing fair transportation systems* (Abingdon: Routledge, 2016), 22.

18. It is my understanding that "heroes" is now a gender-neutral term.

19. Shane Epting, "Urban infrastructure and the problem of moral praise," *Techné: Research in Philosophy and Technology* 25, no. 2 (forthcoming).

20. For more information on choppers, see Kollies Parts, "What's a Bobber and a Chopper?" https://www.kolliesparts.nl/en/blogs/kollies-parts-blog/whats-a-bobber-and-whats-a-chopper/. 2021.

21. For a complete view of structural ethics, see Philip Brey, "From moral agents to moral factors: The structural ethics approach," in *The moral status of technical artefacts*, eds. Peter Kroes and Peter-Pauk Verbeek (Dordrecht, NL: Springer, 2014), 125–42.

22. Brey, "Moral Agents."

23. For an in-depth look at such positions, see Shane Epting, "Urban infrastructure and the problem of moral praise," *Techné: Research in Philosophy and Technology* 25, no. 2 (forthcoming).

24. For a recent examination of this approach, see Shane Epting, "On moral prioritization in environmental ethics: Weak anthropocentrism for the city," *Environmental Ethics* 39, no. 2 (2017): 131–46. One of the founding papers that established weak anthropocentrism robustly is from Eugene Hargrove. For more information, see Eugene Hargrove, "Weak anthropocentric intrinsic value," *The Monist* 75, no. 2 (1992): 183–207.

Chapter 2

Moving and Thinking

I am going to answer this question in a few different ways: What are urban transportation systems? A short, nontechnical response is that they are technologies, mostly for the sake of something else. For instance, while going for a drive or a ride can be for fun or relaxation, typically, trips in an urban setting involve going somewhere for a particular purpose. Implied within this notion is the idea that transportation systems are parts of our daily lives, so much so that they blend in with the city as a backdrop. In turn, we could simply accept them as parts of the stage where our lives play out. It seems as if we have no choice about their placement or design, so why should we bother thinking about them when we can just keep moving, carrying on with our affairs? Once we see the complexity and intimacy that comes with transportation systems, as they help create the "urban condition" that manifests on our city streets, the hope is that this situation motivates us to think more about moving.

With these points in mind, there is a need to situate transportation systems into a broader context. On the one hand, the reason behind this move is to see that there are distinct elements that will help us understand how urban mobility plays into the totality of technological advancement. That is, examining our transportation systems serves as a glimpse into the panorama of the collective human-technology endeavor. This view shows how transportation affects other urban elements, people, the city, and the planet.

On the other hand, by thinking about transportation systems in this manner, we also understand how systems such as governments and global trends (e.g., commerce) will impact and continue to shape urban mobility. We cannot think about transportation systems in isolation from other considerations because larger socio-material arrangements always surround them. Even though we can separate them to obtain a particular view, the concern here is for urban mobility, which is why the connotation of urbanity will always accompany the term "transportation system" in this text. In turn, we must remember that we are dealing with navigating the cityscape and its numerous entailments. This point does not suggest that rural mobility is not of concern, but each topic deserves its own spotlight to give it the attention it deserves.

In turn, this chapter brings Hans Jonas's major work, *The Imperative of Responsibility: In Search of an Ethics for the Technological Age* into view. After exploring the role of his wisdom in technology's future, the focus shifts to an expansive appreciation and slight critique of the insights that he reveals regarding humankind's moral obligations in terms of technology's total impact on the planet and its parts. To apply his thinking in a manner that clarifies the significance of transportation systems, I pay specific attention to cities. This focus is of paramount importance, considering that they play significant roles in determining the quality of people's lives, especially regarding how they access it via mobility. Framing such issues using a Jonasian-inspired framework lets us choose good courses for future action while learning from history's moral shortcomings regarding technology. Examining technology in such a way will provide the necessary background for us to see a full range of considerations for urban mobility.

For instance, in terms of the considerations that suggest why he was making his inquires, Jonas's primary motivation in *Imperative* was to reveal how traditional ethical theories cannot account for the moral consideration of nonhuman species, ecosystems, the accumulating effects of technology, the global impact of technology, and future generations.[1] Along with his progress in thinking through these affairs, continuing his line of thought could yield mitigatory avenues for the problems that modern technology helps cause. Such an exercise aims to lay out the conceptual characteristics that could help improve and develop alternative transportation technologies that reflect better values in the outcomes that they could help achieve. Facilitating such labors includes identifying the thinking that benefits the endeavors such transformation requires.

Unpacking these claims is the goal of this chapter. It begins by laying out Jonas's ideas as mentioned above, exploring them to see how transportation fits into a much larger picture of human-technology relations, which includes a critique of Jonas that can help us continue his thinking for the city. After exploring these aspects, I illustrate the kind of thinking that is advantageous for charting a way to deliver a future that does not resemble the present in terms of harmful outputs.

THE PROMISE OF JONAS'S TECHNOLOGICAL ETHICS

While canonical moral theories can help us address human-to-human affairs, these issues are of little concern to Jonas in *Imperative*. Instead, he is mainly dealing with their limitations, which are revealed when we try to square those theories with the problems associated with modern technology. Although the nature of such affairs has been widely investigated in academia, many of the issues that Jonas was scrutinizing were embryonic when he penned them in the late 1970s, especially in subfields such as environmental ethics that were beginning to take root.

Due to these conditions, Jonas made numerous insightful examinations of the moral shortcomings regarding technology and the nonhuman world, which would appear commonplace among contemporary environmental philosophers.[2] Today, his works are slowly gaining momentum in environmental philosophy, most notably as they intersect with the philosophy of technology and environmental justice.[3] Still, there are several benefits to discovering the myriad ways to employ lessons from his thinking to guide us as we move forward to address the matters he was confronting.

What is hugely significant about the nature of Jonas's enterprise, as mentioned above, is its incredibly broad scope. Consider, for instance, that often theoretical objections in ethics are singular. For example, a common criticism of deontology is that it is too strict and does not allow reasonable exceptions.[4] In terms of utilitarianism, it does not allow people to make decisions t that consider the distinction between persons (e.g., closeness of relations such as family).[5] Instead of raising these kinds of isolated objections, Jonas attacks the entire canon for its inherent limitations when it comes to technology. From the outset in *Imperative*, he begins with a bold opening that sets the stage for a criticism that takes over 200 pages to unpack fully:

> All previous ethics—whether in the form of issuing direct enjoinders to do and not to do certain things, or in the form of defining principles for such enjoinders, or in the form of establishing the ground of obligation for obeying such principles—had these interconnected tacit premises in common: that the human condition, determined by the nature of man and the nature of things, was given once for all; that the human good on that basis was readily determinable; and that the range of human action and therefore responsibility was narrowly circumscribed. It will be the burden of my argument to show that these premises no longer hold, and to reflect on the meaning of this fact for our moral condition.[6]

To put this criticism in a perspective that shows the depth of its significance, if we take his thinking on these matters seriously, we must also ready ourselves to receive an expanded scope of responsibility. This notion entails that we must not only revisit the structure of the history of ethical theory, but we must also broaden how we conceive of what we expect it to do for us. Speaking meta-ethically, what can we reasonably set as a goal for ethical inquiry once we attend to the matters that Jonas raises above? Considering that he is urging us to look at his work as an *imperative*, we cannot discount the sense of urgency that accompanies this term. One way to approach this topic is to think that giving it sufficient dedication means that we must attend to the question in a way that seeks to provide an answer that matches its intensity concerning the boldness of its scope.

For instance, Jonas continues to flesh out the entailments that his inquiry brings with it—ones that challenge humankind to robustly examine the new conditions that we find with modern technology:

> The changed nature of human action calls for a change in ethics as well: this not merely in the sense that new objects of action have added to the case material on which received rules of conduct are to be applied, but in the more radical sense that the qualitatively novel nature of certain of our actions has opened up a whole new dimension of ethical relevance for which there is no precedent in the standards and canons of traditional ethics. The novel powers I have in mind are, of course, those of modern technology.[7]

This passage is significant because of the claim that traditional ethics cannot deal with the problems that humankind has created with technology. This point might appear trivial on the surface. However, such a view is shortsighted. Consider, for instance, that one could make a case that if inventors and engineers were using the canon of ethical thought to guide the thinking that led to our planet's current state, then there would not be so many issues that stem from their works. Yet, that position does not hold.

Countless contemporary environmental conditions such as sea-level rise, deforestation, and mass extinction show otherwise. Due to such outcomes, Jonas's thoughts suggest that we have to question our ethical systems' orientation from their foundations when it comes to technology. When we do, we find that they were ill-prepared from the outset. They were not equipped to deal with the problems that modern technology has caused. For this reason, calling the canon into question becomes paramount for Jonas.

In turn, we need to examine the strengths and weaknesses of traditional morality as employed in the literature and the classroom. On the one hand, we cannot only advocate for the protection of wetlands, baby seals, or trees, as several extensionists have done, speaking broadly. Such efforts merely extend the application of thought systems that Jonas has already shown lack the necessary longevity and integrity to help us make moral sense of technology. The point is not that we must attend to ecological problems as they appear. Such an approach would not get at the heart of the matter, leaving us unsatisfied with the need to address technological concerns.

On the other hand, we must not wholly scrap the canon. We must reformulate it so that it can help us in the endeavor to develop a system of thinking that lends itself to creating a world that lacks the harmful characteristics that have pushed humankind into a corner. It is one wherein we must rethink our most significant formulations of morality. We can square this point with the canon because, in several instances throughout *Imperative*, Jonas employs components of ethical theories as theoretical tools that can facilitate the grand undertaking that he envisions can save us from ourselves .[8]

In this sense, one could make a case that he, too, was building an "anti-framework" that could tend to the issues of such a grand scale that the problems with the technical congruence of the interplay of theories were trivial. One way to think about it is as follows: we can repurpose tenets of moral theories to help us address technological, moral problems—after we have re-examined and rebuilt our foundation. While this idea is consistent with Jonas's use of the phrase "radical sense" above, the more appropriate use of the term "radical" is to go as far as possible with restructuring the foundation of our thoughts on such matters. This point requires focusing on a usage that maximizes the utility of its meaning.

For example, Jonas mentions that the effects of technology have created the conditions wherein moral theory lacks precedent, suggesting that such traditional frameworks are inherently limited to deal with these kinds of issues (e.g., climate change). In turn, we must think and act so that we can advance our moral thinking in a manner that can keep pace with the evolving character of technological problems, especially as they pertain to affairs in urban mobility. As I had mentioned in chapter 1, these circumstances require an approach that is flexible, meaning that it can adapt to new elements. This factor of adaptivity is missing from most canonical moral theories, at least in the sense that it was explicit and underscored.[9]

If it would have been part of ethics, then Jonas could have argued that such changes could have been made to those theories. This point exhibits that he would not have had to challenge "all previous ethics" from the outset of *Imperative*.[10] However, considering that he was holding that their designs were limiting, humankind must now learn to think in a manner that pushes back against this reality. In turn, we must apply this new kind of moral thinking to all of our modern technological affairs, a feat that is quite extensive. Yet, for transportation, we can narrow the focus to a technological area with significant importance for urban living.

Throughout the book, for instance, he takes on an intensely broad range of issues to illustrate how many of them come into play when trying to develop the necessary framework that can guide our thinking on the ethics of technology: socialism, capitalism, science, and democracy—just to name a few.[11] Although engaging in these issues falls far outside of this chapter's topic, they all relate to the limited moral lens that needs switching to fully see the panorama of problems that modern technology has caused, which now require us to gain new ground for our moral footing. On the foundational level, we can utilize his thinking to illustrate that the structure of our moral outlook needs a retooling to approach the issues mentioned above adequately.

For example, the criticisms that Jonas provides show that canonical moral theories qualify as strongly anthropocentric.[12] While this issue gestures toward basic moral extensionism, that quality is merely a coextensive feature that is also required for establishing a solid foundation for our ethical dealings with modern technology and the nonhuman world. Yet, writing long before terms such as ecocentrism were common among environmental philosophers, Jonas was appealing to ecosystems and nonhuman species' intrinsic value. By appealing to this aspect as a prime element that the ethical canon lacked, Jonas arrives at a cornerstone of the "extrahuman" thinking necessary to develop the attitude to rebuild our moral systems to deal with modern technology.[13]

With this criticism in place, he illustrates that through lacking this element in the canon of moral thought, we could not address the issues of time and the accumulating effects of technology on the nonhuman world.[14] While these affairs brought nonhumans into view, they also brought the idea that there is also a global element that we must consider.[15] That is, technological actions are not only local, but they have far-reaching impacts that climate change makes evident. In turn, these issues could impact people in the future, a notion that was speaking to calls for sustainability long before the term became a staple of the academic lexicon.

Bearing the above notions in mind, the foundation that Jonas seeks to build is daunting. To establish a way to deal with the expansive range of problems that we find with modern technology, along with its effects that are not immediately obvious, he gives us an incredibly vague "imperative of responsibility." It is expected to guide us in our technological dealings: "Act so that the effects of your actions are compatible with the permeance of genuine human life."[16] While this dictum is guilty of being incredibly imprecise, such criticism misses the point that he does not make an argument per se. He is expressing wisdom. The wisdom here is that, while we are wrestling with how we are to employ technology to live for the long term, we must also use this guiding principle in all of our technological affairs for the short term that will get us there, hopefully. To push back against this point, it is doubtful that Jonas could have anticipated the exact remedy required to deal with specific technological problems, ones that will change with the circumstances and remain particularly suited for each city. Yet, the point that deserves underscoring is that he spoke to considerations that go beyond individual issues in technology and ethics.

Jonas was formulating a response that would transcend the immediate, one that would serve our indefinite yet limited existence on this planet. While academic criticisms are arguably not in short supply, one can hold that wisdom is not a commodity or something that is easy to come by. In this case, making sense of the whole and developing an encompassing guide forward remains widely applicable to changing times. Jonas was serving us well. To use his words: "We need wisdom the most when we believe in it the least."[17] Applying such insights to the case at hand, we gain a position from where we can examine how transportation systems fit in with the rest of the city, which includes but is not limited to infrastructure, commerce, recreation, health, and education.

Due to his approach's inherent characteristics, we can further build on Jonas's thinking when moving forward, making mitigatory efforts and working toward solutions to the moral problems that we have inadvertently created collectively with technology. We can focus on transportation technologies as they relate to and affect the areas mentioned above. In turn, one can argue that the robust character of Jonas's challenge for traditional moral theory is humbling, considering how it can serve not only its capacity to be widely applicable. Yet, it also provides us with a guiding spirit, one that reminds us why we care about urban mobility in the first place.

To engage with his thought meaningfully entails that we must address the expansive array of elements that now must occupy space in our thinking as we craft a way forward in our approaches to such problems. Keeping in mind that Jonas is not merely thinking about expanding the entities that deserve inclusion into our ethical purview, but he is thinking on a much broader scale that requires us to examine the foundations of how we conceptualize such affairs. This idea suggests that once we grasp the nature of his imperative for technology, we must situate ourselves so that we can not only reevaluate the scope of ethical inclusion, but we must also anticipate what thinking about ethics in such a manner includes. If we are going to "radically" reconceive the foundations of moral inquiry for the sake of modern technology, then one way to follow this view is to hold that we must do the same for how we respond to them.

Although the academic norm is to engage in endless debates, the fact that we are dealing with a topic that warrants an imperative *for* technology could mean that we go beyond our traditional boundaries, participating in interdisciplinary and trans-disciplinary conversations. In the section below, I explore these topics, aligning them with select insights that neighboring researchers have pinpointed to help create better conditions for the city and its mobility networks. Through thinking with these notions in concert, we can better work toward dealing with the totality of our technological condition as it appears in our cityscapes.

TOWARD TECHNOLOGIES FOR THE CITY

Let us consider that a radical restructuring of ethics *could* guide technology, focusing on urban elements. If this were the case, we should expect that the city's future might not resemble its past in terms of harmful outputs. A critical factor that must come into view here that will help us understand the far-reaching effects that concerned Jonas was that utopian-inspired sentiments accompanied a "received view" of technology.[18] This idea suggests that humankind sought to advance technology to achieve this goal.

Despite the veracity of this position historically, it did not materialize so that we could engage in activities that would help us flourish without much worry. Those points aside, this connotation must be disassociated from our technological pursuits. We need not satisfy Jonas's view on this point to pay a debt to him. Yet, a utopian view of technology serves nary a practical interest going forward. Developing technologies that do not exacerbate or perpetuate the harms mentioned earlier, that motivated Jonas to investigate previous forms of ethics, deserve the pursuit for their ability to produce the mitigatory outcomes that we desire.

Bearing this point in mind, if we do side with Jonas's anti-utopian commitment, then we can continue to shape a new conceptualization of progress in technology or amend a previous one. In turn, if we seriously consider Jonas's insights, then our thinking on technology not only calls for an ethical re-grounding, but it also requires that we shift our mindset with regard to the spirit that is to guide its pursuit. By coupling these supporting theoretical elements, reconfiguring a view of technology that at least aims to deliver less harmful outcomes has a chance to find success, which could include abandoning the position that technology should support utopian visions.

Through developing a conception of technology that does not include the aspects mentioned above, the focus can shift to developing an attitude toward technology that can mitigate social and ecological harm. Such notions suggest that the technologies that would support such a bold endeavor might not hold the same characteristics that we recognize today. These points should concern philosophers due to their normative dimensions. They should also be of interest to engineers, planners, architects, and other professionals who will help reshape the city's technological dimensions with their ingenuity. Together, these approaches can put Jonas's ideas in motion, securing them so that we can have a future worth wanting.

To develop a manner of thinking that can guide this point, we must seriously consider that almost every facet of urban life requires attention. Jonas identifies several focus areas that can provide a reorientation to technology to assist in such an undertaking. His wisdom regarding these topics will significantly enhance efforts to understand how technologies such as transportation networks, in terms of wholes and their associated specific parts, have produced drastic outcomes that warrant an ethical inquiry, as mentioned earlier. A beneficial way to situate urban mobility for this task is to think about it as dependent on technological parts. One that must be thought about in relation to other social and technical parts—one that remains external to transportation wholes—is to see how Jonas thinks about the city. We can then examine how he fleshed out the role of urban systems such as transportation in a much broader context to gain an enhanced view of transportation's relationship with the city.

For instance, Jonas tacitly employs the concept of the city as a theoretical device at the beginning of *Imperative*.[19] He stacks the city against nature to juxtapose ethics before and after modern technology. For example, he refers to humankind's home as the man-made" island of the city, pointing out that the balance between humankind's habitat and the nonhuman world does not hold steady any longer.[20] In a Jonasian-inspired view, one can argue that cities are anthropogenic artifacts—technologies.[21] We can unpack this view to mean that cities are "devices" that we use to accomplish specific tasks, including living, working, playing, and discovering, along with all the other things that make life in the city worth attaining. In short, they are tools for worthwhile goals such as socially just urban sustainability or human flourishing—or at least such outcomes could be the case.

Stacking modern cities against ancient ones, they could be thought about as categorically different, although they were used to accomplish the same goals as above. They would still count as being technologies, but modern cities significantly differ due to their impacts. This point does not assume that ancient cities were without fault. Yet, it holds that modern cities, due to aspects such as carbon outputs and massive land consumption, have effects that we did not see readily with ancient cities. The purpose of thinking of the city in this manner is to exhibit how cities have affected nonhuman nature. For instance, Jonas holds that:

> For the boundary between "city" and "nature" has been obliterated: the city of men, once an enclave in the nonhuman world, spreads over the whole terrestrial nature and usurps its place. The difference between the artificial and the natural has vanished, the natural is swallowed up in the sphere of the artificial, and at the same time the total (the works of man that have become "the world" and as such envelop their markers) generates a "nature" of its own, that is, a necessity with which human freedom has to cope in an entirely new sense.[22]

In the passage above, Jonas makes this juxtaposition to elucidate the magnitude of how modern technology has drastically impacted the world. Now we have to address these concerns in every aspect of urban life that requires natural resources. Consider, for instance, that when it comes to urban spaces, we can point to the wide-scale use of natural resources to accommodate urban sprawl and highways.[23]

For example, if we consider that the vast majority of transportation systems in the United States rely on vehicles with gasoline-powered engines, then it should not come as a surprise that they are the largest producers of pollutants that are responsible for harmful elements such as smog and acid rain, along with how they affect the tropospheric ozone.[24] In turn, this reliance perpetuates a situation wherein these technologies' harmful effects remain constant and increasingly difficult to mitigate.

In turn, as technologies, modern cities qualify as instruments that contribute to the inequality of environmental harms and benefits. Jonas recognizes several of the areas that do require a rethinking, holding that we need to pay attention to food production, raw materials, and energy.[25] Even though the list above is brief, it indicates the kind of changes required to address the demands embedded within his imperative. Bound to these ideas is the more extensive consideration that our collective technological pursuits carry a connotation that we are engaged in such enterprises because we, in the sense of a "received view," expect technology to support utopian visions. Such quests are pushing nature to its limits. In turn, we must ask: Where are the limits of what we can expect from nonhuman nature? To unpack such matters, Jonas argues that:

> The question as a whole lies in the domain of the infant science of ecology, and as the particulars in the fields of the biologist, the agronomist, and so forth. In addition, it also ropes in the economist and engineer, the city planner and transportation specialist, and so forth. Only the interdisciplinary pooling and integration of all these will lead to the global environmental science that is needed. Here the philosopher has nothing to say, only to listen.[26]

Bearing the above passage in mind while addressing concerns for the city, the practical problems about how we are to develop a manner of living consistent with Jonas's imperative will require advanced technical expertise, which most philosophers are unqualified to address. Yet, they can contribute to such an undertaking in several ways. One such way that philosophers can benefit the complex, wicked problems that we are facing is by paying attention to urban technologies' moral dimensions, zeroing in on areas of concern, including transportation.

While Jonas was speaking to broad concerns through wisdom, we can couple his work with other kinds of advancements to move our thinking on such affairs forward. For example, based on their research showing that growing cities follow basic principles that remain identifiable in several cities across the globe, two theoretical physicists who have an interdisciplinary orientation, Luis Bettencourt and Geoffrey West, make a bold call for researchers and professionals to contribute to a unified theory" of sustainability for urban environments.[27] Their studies creatively develop a useful science of the city," which could serve municipal professionals who are contending with problems that arise from population increases.[28] Their work clearly shows how examining the structural conceptions of cities can have meaningful, significant advantages that can help municipal workers address emerging challenges that remain inherent concerning how cities increase in population.

These researchers illustrate that many expanding urban centers take on an element of predictability, which suggests that there are underlying principles at play. We can extrapolate details about the specific nature of urban growth. Despite numerous kinds of change and uncertainty surrounding cities, they correlate to an essential, albeit not exact, pattern. It provides a way to anticipate the required or saved resources that a city will "consume." This research illustrates that when cities double their numbers, they only need an additional 85 percent increase in associated infrastructure in terms of elements such as roads and water resources.[29] Through developing this data and its suggestive conclusions, Bettencourt and West exhibit that cities that are built for high density use resources more efficiently than smaller ones.[30]

Due to this pattern, larger metropolitan environments do more to support Jonas's view of bringing our disciplinary lenses together to gain the panorama that shows how to support sustainability that benefits our long-term ability to survive and, hopefully, thrive.[31] This point is significant because it shows that we can develop a methodology to study cities to derive principles that will support anticipatory measures. The advantage here is that we gain the means to make predictions about how actions such as implementing smart meters, adding bicycle lanes, or upgrading infrastructures and services can help decrease carbon outputs and other unwanted outcomes as they notice population increases. Bettencourt and West's research is arguably in its developing stages when compared to other areas of study. Yet, there is promise in the idea of continuing this line of work alongside other disciplines—including philosophy.

These interdisciplinary and concerted efforts can help move us in the direction that Jonas champions, the "global environmental science" that he claims is needed to remedy technology's ill effects.[32] The city provides an approachable and manageable way for us to engage in such undertakings. When we advance such efforts, we can begin to deal with the criticism that I raised against Jonas earlier that concerns his imperative's vagueness.

The first step to pursuing such a practice is to determine where to direct the needed efforts, a plan of action determined by the amount and kind of required attention. Recalling the passage above, philosophers working on topics of relevant concerns depend on interdisciplinary researchers who can illustrate the real-world problems using their disciplinary strengths to provide evidence. Consider, for instance, that Jonas, as a philosopher, listed what he thought were the most pressing technological issues that needed attention. He claimed that raw materials, agriculture, and energy were the primary areas that need investigations due to their impacts on the planet.[33] Although these topics are of paramount importance, they might not be the most pressing concerns, despite the overwhelming amount of good that addressing these issues could bring.

Consider, for instance, that Jonas does not provide any of the needed empirical data to bolster his argument. Viewed in this way, advocating for these specific areas, without data, is guesswork in the best case and arbitrary and ill-founded or harmful in the worst one. While there is little doubt that attending to these issues, making them a priority for research would help us secure the conditions for humankind's permeance for which Jonas champions, advancing such a research agenda requires that we address two issues.

First, Jonas assumes that addressing agriculture, energy, and securing raw materials is the best or most efficient way to decrease our demands of the nonhuman world. Providing evidence of this situation is best suited for the professionals who study fields such as industrial ecology and urban ecology, considering that this work keeps track of materials flowing in and out of urban places and how human-centered elements intersect with nonhuman life.[34] When we look at these kinds of numbers, we discover that cities consume between 60 and 80 percent of the world's energy, 75 percent of the resources that come from the nonhuman world, and account for 75 percent of greenhouse gas emissions.[35] Along with this sobering reality, since the 1970s, greenhouse gas emissions from all modes of mobility on land, in air, and on water have doubled.[36] Vehicles on roadways account for about 80 percent of this increase.[37]

Yet, these conditions' actualities should not be the primary focus when confronting the problems associated with urban mobility, despite having a monumental significance for the planet. That is, the element that is missing in Jonas's assessment is how modern technology—on the grand scales that he introduces (e.g., raw materials, agriculture, and energy)—concerns how it impacts the people who struggle the most. These are the people dealing with greater degrees of injustice and are burdened disproportionately. These people are the ones who would benefit the most from a moral assessment of technology, but they are absent from his views in *Imperative*. What is required, then, is a critique of Jonas's position that brings these notions into view. Due to this situation, before moving forward with any substantial technological renovations of the urban sphere, this point will require significant attention.

While we can prioritize who is to receive the initial benefits of retooling transportation systems due to a moral bearing, interdisciplinary researchers provide us with significant motivations to act in such a manner without philosophic appeals. That is, based on data collected, the primary reason why people who are stuck in poverty remain there is due to the many challenges of transportation.[38] This point does not suggest that the other areas Jonas brings up are not worthy of investigation. However, when addressing transportation alongside the categories that Jonas mentions, there is no reason why it could not be included in such a list, especially considering its significant role in determining the quality of people's lives who are trapped in poverty.[39]

Although all of the issues mentioned above come from studies that depend on researchers who are housed in several different areas of the academy, they share common ground in that they help us understand the urban condition as it relates to urban mobility. Each of them is engaged in assessing outcomes that stem from current sociopolitical arrangements that manifest as problems on city streets, boulevards, and highways. This notion means that each researcher is engaged in evaluating outcomes, but they are not in the business of morally discussing the conditions that yield troubling situations, and of course, they are not directing or recommending ethical ways forward. However, other research in the academy can inform the specific manner wherein we move on to approach the problems that plague us from the technological side of affairs. For such a task, philosophical enterprises can be of great benefit due to the ways that they show how categories differ, how to define concepts methodically, and yield insights on subjects such as values.

Aside from this point, understanding these topics from outside engineering and planning can benefit how we view the complex character of mobility's moral dimensions. In turn, chapter 3 moves toward a way to orient the conversation toward dealing with transportation outcomes through examining it on its most fundamental level. Recalling the inquiry from this chapter's outset, "What is an urban transportation system?" I answer this question in a manner that displays a wide range of possibilities that moves far beyond a technical description.

NOTES

1. Hans Jonas, *The imperative of responsibility: In search of an ethics for the technological age* (Chicago: University of Chicago Press, 1984), 4*ff*.

2. Shane Epting, "Questioning technology's role in environmental ethics: Weak anthropocentrism revisited," *Interdisciplinary Environmental Review* 11, no. 1 (2010): 20.

3. For a recent example of this work, see Jérôme Ballet *et al.*, "Hans Jonas: Bridging the gap between environmental justice and environmental ethics," *Environmental Ethics* 39, no. 2 (2017): 175–91.

4. Larry Alexander and Michael Moore, "Deontological Ethics," *The Stanford Encyclopedia of Philosophy* (Winter 2020 Edition), eds. E. Zalta, https://plato.stanford.edu/archives/win2020/entries/ethics-deontological/.

5. John Rawls, *A theory of justice,* (Cambridge, MA: Harvard University Press, 1999) 24.

6. Jonas, *Imperative*, 1.

7. Jonas, *Imperative*, 1.

8. Two examples of how Jonas appeals to canonical theories to serve as conceptual devices come to mind that support the claim is found in how he uses deontology and virtue ethics. For instance, when considering the significance of actions required to deliver humankind from the kind of technological thinking that has created conditions such as anthropogenic environmental degradation, Jonas appeals to the structure of deontology to develop his moral imperative of technology. Second, when he is dealing with the possibility of being accused of adopting an anti-technological position, he makes use of virtue ethics, saying that humankind has weighed too heavily on hope for technology. Balancing our thinking now requires us to spend some time on the opposite end of that scale, which entails acknowledging and accepting the fear that ignoring such an imperative could bring. To see the former appeal, see Jonas, *Imperative*, 1. To see how he employs the latter, see Jonas, *Imperative*, 204.

9. One could argue that deontology's structure meets this condition, considering that it makes room for absolutes, which entail that it is equipped to address any such new problem. While this point is formidable, it merely avoids the reality that deontology remains inflexible (without modification). It forces the world to conform to it, rather than bending in a manner that makes it accessible, aligning with Lewis Gordon's notion of disciplinary decadence. Appealing to deontology assumes that we want a moral theory in the first place. However, these issues do fall outside the scope of this project.

10. Jonas, *Imperative*, 1.

11. Ibid., 142*ff.*

12. Ibid., 4.

13. Ibid., 8.

14. Ibid., 4–6.

15. Ibid., 8.

16. Ibid., 11.

17. Ibid., 21.

18. For an understanding of how Jonas situates technology and utopia, see Jonas, *Imperative*, 176*ff* and 201*ff.*

19. Ibid., 3.

20. Ibid.

21. I make a similar point here: Shane Epting, “On municipalities as technologies,” *Philosophy & Technology* (2021): 1–11. https://doi.org/10.1007/s13347-020-00438-z.

22. Jonas, *Imperative*, 10.

23. Ernest Hennig *et al.*, “Multi-scale analysis of urban sprawl in Europe: Towards a European de-sprawling strategy,” *Land Use Policy* 49 (2015): 483.

24. “About Smog, Soot, and Other Air Pollution from Transportation,” *Smog, Soot, and Other Air Pollution from Transportation*, United States Environmental Protection Agency, last updated March 18, 2019. https://www.epa.gov/transportation-air-pollution-and-climate-change/smog-soot-and-local-air-pollution.

25. It is worth mentioning that this task is deeply embedded within the context of Jonas’ critique of the utopian lens that has remained associated with technological pursuits, especially modern technology. However, his insights support his overall endeavor as examined thus far in the chapter and hold steady as part of his critique of the ethics of technology. For more information on this topic, see Jonas, *Imperative*, 178*ff*.

26. Jonas, *Imperative*, 189.

27. Luis Bettencourt *et al.*, “A unified theory of urban living,” *Nature* 467, no. 7318 (2010): 912–13.

28. Luis Bettencourt *et al.*, “A unified theory,” 912–13.

29. Ibid.

30. Ibid.

31. It is worth pointing out that cities vary considerably in terms of their resource requirements, which will challenge this or any other emerging paradigm. Still, Bettencourt and West hold that cities continue to expand their populations. The space required for living decreases due to elements such as density, economic outputs increase, and cultural dimensions takes root. The downside is that as cities continue to grow and improve in these ways, unfortunately, problems such as crime, traffic, and health issues also increase along the same lines. While such a reality does exist, however, knowing about these affairs could help municipal officials to tackle these problems in a way that lets them measure the effectiveness of their chosen measures.

32. Jonas, *Imperative*, 189.

33. Ibid.

34. Xuemei Bai *et al.*, “Urban ecology and industrial ecology,” in *Handbook of urban ecology*, eds. D. Ian, D. Goode, M. Houck and R. Wang (2010), 26. It is also worth mentioning the Xuemei Bai has recently argued that a bridge between urban energy and material flows can provide scientific insights for urban professionals who are working on related affairs. However, one could say that such analysis would also benefit from ontological and/or ethical assessments to determine routes to improved efficiency and normative evaluations, respectively. For more information, see Xuemei Bai, “Eight energy and material flow characteristics of urban ecosystems,” *Ambio* 45, no. 7 (2016): 819–30.

35. Achim Steiner, “Forward,” in *United Nations environment programme* (2013), 4.

36. Ralph Sims *et al.*, “Transport climate change 2014: Mitigation of climate change,” in *Contribution of working group III to the fifth assessment report of the*

intergovernmental panel on climate change, eds. O Edenhofer et al. (Cambridge and New York: Cambridge University Press, 2014), 605.

37. Sims, "Transport," 605.

38. Evelyn Blumenberg *et al.*, "Getting around when you're just getting by: Transportation survival strategies of the poor," *Journal of Poverty* 18, no. 4 (2014): 356. Mikayla Bouchard, "Transportation emerges as crucial to escaping poverty," *The New York Times*, May 7, 2015.

39. Evelyn Blumenberg *et al.*, "Getting around," 356. Bouchard, "Transportation," *The New York Times*, May 7, 2015.

Chapter 3

Thinking, Moving, and Parts

Urban mobility surrounds us. Yet, one could argue that many people hardly notice it, even though it helps shape the quality of our lives the world over. Although some cities fare better than other places in this regard, their successes and shortcomings serve as a way to compare and contrast what works and what does not. For most researchers and professionals who deal with transportation systems, even though their definitions vary, what "works" arguably has an understood meaning that provides the common ground for conversation. It means something along these lines: "enables people to go and return from destinations within a given city or region."[1] In philosophical terms per the academy's standards, however, its disciplinary orientation demands that we scrutinize such groundings to see if they will collapse or hold steady, allowing us to build elaborate systems of thought.

Examining and debating the meaning of such a term could go on indefinitely. Nevertheless, transportation involves ethical issues that are always beyond, yet remain subject to, the theoretical realm. Real people's lives and livelihoods are at the center of urban mobility—and they are at stake in every decision that includes aspects such as infrastructure, policies, and budgets. The matters at hand bring the interests of all urban and suburban dwellers into the picture. Yet, the degree that different groups are affected, positively and negatively, indicates the numerous social ills that have worn the fabric of many societies too thin in some instances while staining it during darker times.

Such claims do not require that we name any specific city or region. It is unlikely that there is one that does not have any transportation issues, especially when considering global climate change. That is, partly due to fossil-fuel consumption, it impacts humans, nonhumans, and these new global conditions will continue to affect people and nonhuman nature well into the distant future. In turn, transportation challenges require solutions beyond the thinking that comes with humankind's immediate mobility needs.

This point does not suggest that some municipalities or regions have not discovered transportation solutions that produce less harmful, more equitable, or more liberating outcomes. Many places are leading the way in mitigating harm and developing paths to an improved future for urban mobility—one bike lane or bus-rapid-transit network at a time. Arguably, several local transport authorities generally care and act to improve their residents' lives, often focusing intensely on the people contending with forces and circumstances that are mostly if not entirely beyond their control.[2]

How do we know which communities are being neglected, oppressed, dismissed, or diminished—to the extent that demands immediate attention—when considering outcomes that transportation systems help produce? The answer depends on the city or region in question, and asking the people who live there and are facing the dire conditions mentioned above *should* (in the normative sense) inform the answer. This idea necessitates that if such people are not meaningfully included in the relevant conversations that pertain to transportation systems, then leaving them out is a separate but related moral issue. In addition to this idea, municipalities and regions have unique histories that compound problems in highly distinct fashions. The people involved in these histories need to weigh in when it comes to affairs in urban mobility.

Bearing this point in mind, we could be dealing with long-standing issues that go back generations in some cities, often intertwined with extremely upsetting acts of violence that require looking at the larger context of urban planning's history in a given area. While one could argue that we still have to deal with problems on a case-by-case basis when charting a way forward, only addressing contemporary issues fails to consider transportation's dreadful past in some cases.[3] Lurking behind these incidents are unasked questions that concern their long-term psychological, economic, social, cultural, and political elements tied to those shameful actions, not to mention amends-making. How do we square such actualities with the needs of the current realities? Giving these inquiries their due cannot be answered in a monograph. They need research centers at a minimum.

Aside from this point, transportation planners are tasked with dealing with anticipated mobility needs. Yet, as Karel Martens argues, marching forward while focusing almost exclusively on this prerogative is dangerous.[4] It would allow for the perpetuation of existing injustice, which would continue to cause harm and pass up opportunities to *begin* to touch on righting history's wrongs. Bearing this point in mind, transportation professionals might see highway expansions as mere ways to ease congestion, but such an undertaking is hardly straightforward in some instances. Despite this situation, it seems incredibly challenging, in a practical sense, to give such topics the respect they deserve while aiming to make the trains run on time. This subject is above the pay grade for a philosopher, but it does concern the prioritization of values. So, there is that.

In dealing with this dilemma, however, we must keep in mind that the people who will suffer from increasingly poor air quality, island-heat effect, and dangerous decibel outputs, and those who could have had or will have their lands, homes, and businesses taken for land acquisition (e.g., expropriation, compulsory purchase, eminent domain) in the "public's" interest might disagree. They could hold that such acts, such as adding a traffic lane, while aiming to improve urban mobility, are nothing more than a continuation of harmful actions that go back to the evils of the slave trade in some places. I cannot—morally or epistemologically—speak for them—for what they need for justice, or the practical realities of mobility.

Nevertheless, I have no problem saying that denying people the right to have a *meaningful* voice in the decisions that will impact their lives is something that a person with power should not do. Although such concerns will vary from one place to the next, such histories exist in numerous contexts, meaning that these considerations must hold steady as paramount considerations when planning and maintaining transportation systems. This notion suggests that gaining intimate input on how individual parts that belong to transportation systems and more complex groupings of parts will require their voices to determine the processes that will help shape urban mobility. With this point in mind, including such groups into future decisions, coupled with the idea of previous wrongs that have been committed against them, we see that, for several cities, transportation planning should support reparative efforts. Such an approach entails that the people who have been harmed might need to discuss the future of mobility with the specialists who will design transportation networks to contribute to charting a just path forward.

Although this idea should lean toward universal consideration, no two cities are the same. This reality entails that each urban area will require mitigatory efforts that speak to their specific mobility problems. The groups of people who are subject to socially ingrained oppression vary from one area to the next. Every city has people who need services to help them navigate the cityscape. Their lives matter, and we are in this together. With this point in mind, we must also recall that transportation serves all urban dwellers, suburban workers, and visitors.

No one person or group has the necessary perspective to provide all of the information that is required to attain transportation justice—if it is even possible in the first place. Transportation professionals do not have all of the answers. They cannot gain the necessary information unless they have intimate knowledge about how different groups are affected by the transportation systems that serve them.[5] Yet, those people do not have the technical education that helps the trains run on time. Thinking in concert, however, we can work toward such a conception. Of course, getting into action demands that we examine the realities that remain on the periphery of mainstream urban mobility. Consider the notions below.

Older adults are dying alone. Many of them are unable to leave their homes because they can no longer drive safely.[6] They are cut off from the world. This situation is becoming a public health issue that presents unprecedented challenges. Disabled individuals face discrimination in several forms of public transport. In many instances, these persons lack reasonable transport alternatives.[7] Women and members of the LGBTQIA+ community are harassed and/or assaulted on public transport.[8] Bicyclists and pedestrians face increasingly dangerous conditions.[9] Countless people across the world must fight traffic, and they must deal with its various ill effects. Other people are crammed into train cars, a form of collective hell. Poor people cannot escape poverty because they do not have good mobility options, spending half a workday or more going to a job(s) and back home.[10]

This topic brings up numerous issues. For instance, how are people supposed to advance in life if they do not have the time to engage in activities such as attaining the required education if transport conditions make it so incredibly challenging to attend to basic survival? What do such realities say about the societies that these people belong to? Can such mobility systems push them toward second-class citizenship? While these topics are mostly realities for working adults, there are other issues for children.

In select regions, for instance, children of color grow up with asthma because they live near busy highways.[11] Some highways destroyed neighborhoods that once belonged to marginalized communities, perpetuating systematic racism.[12] How do these children access cultural elements if such necessities remain imperiled? The point here is not to vilify highways. That is, not to put the spotlight entirely on highways as harmful parts of mobility networks, light rail systems achieved the same outcome in certain instances.[13]

Aside from the above anthropocentric concerns, the nonhuman world has received countless harms and destruction. For instance, partly due to the demand for fossil fuels and their associated outputs, the climate is changing, and sea levels are rising, which affect all species. Hundreds of million of nonhuman animals die on roadways each year.[14] Urban expansion also forces them out of their habitats. While these issues are alarming, they only provide a glimpse at the problematic panorama that is many of the world's transportation systems. Moreover, these are the problems that exist today, and people will probably have to deal with worse conditions in the future—unless we do something about it.

Here is the good news: we can. The work here moves in that direction by examining questions and offering answers that create possibilities for alternatives that move in new directions. The trick is to ask the right kinds of questions, even at the most basic level. Yet, the questions must be situated to receive answers that can adequately inform us. The goal is to illustrate how we can reorient our thinking on urban mobility to help us arrive at a future that does not resemble the present.

Another mobility world is possible for a future that, with advanced thinking and some elbow grease, could help people go beyond surviving, but they could thrive in the city. Asking the right questions is indispensable for such an undertaking. For instance, asking the question, "What is a transportation system?" in a shallow sense can only touch on kinds of components and operational functions. This conception is too narrow, meaning that it cannot conceptually reveal structural foundations. In turn, such descriptions cannot help us understand them on a theoretical level. For that task, we must move beyond those limitations. While a transportation system manifests as an entity, the term "transportation system" is also a conceptual device. It provides a way to think about how all of the parts fit together to facilitate urban mobility. Yet, this notion still does not provide us with a robust answer to the question above.

This point indicates that I aspire to dig as deep as possible to arrive at a foundational position about transportation systems. It will provide a way to build a structure that will stand in place once its scaffolding falls away. During this process, one goal is to reveal obscured layers of urban mobility to see many of the elements required for the kind of urban mobility that is worth wanting. Anyone who is concerned with transportation affairs should care about this topic. This view can facilitate a comprehensive understanding of transportation systems that pays attention to its ethical dimensions and their role in shaping the quality of life, as mentioned above. Yet, more than that notion, they can help save us from the world that currently presents us with numerous challenges. It can help us gear transportation systems toward worthwhile goals such as socially just urban sustainability. As a stage-setting endeavor, this chapter starts to unpack these points.

In turn, we discover more elements that help answer the question: "What is a transportation system?" In addition to identifying it as a conceptual device, this chapter pays attention to its ontological structure. Beginning such an enterprise requires taking inventory of what is involved in a transportation system's design and functionality—beyond the surface. To gain this perspective, I want to understand transportation systems as wholes that have numerous smaller parts. The point here is to talk about the individual parts, along with how they can also be parts of larger parts in some cases, which are also parts of the transportation system as a whole. I want to study parts, especially how they relate to other parts to produce the outcomes mentioned above. The catch is that almost anything can be a part, including but not limited to concrete, material parts such as vehicles, trains, and scooters. Yet, there are also abstract parts such as traffic laws, city ordinances, and building codes. Conducting a meaningful analysis will require that we study the parts mentioned above, paying attention to how they fit and interact with each other.

For instance, when thinking about the whole of a transportation system, we can see that there are individual parts that make larger parts. For example, train cars, tracks, and conductors are parts of transport operations. Those entities, when combined, make up a larger part, a rail network. Bicycle lanes, bicycles, and other related parts follow the same pattern. The transport system and the pedal-powered complex are also parts of the larger transportation system as a whole.

The area of philosophy that deals with parts and parthood relations is called *mereology*. Its primary focus deals with how some parts relate to other parts, how parts are part of larger parts, and how they are parts of wholes. Yet, most philosophers who research in this area study mereological topics in highly abstract terms, and my application is "applied" in the best, most generous, sense of that word. This point aside, considering the notions above concerning transportation systems and their composition, mereology easily lends itself to how we can talk about the structure of urban mobility.

In turn, the purpose of introducing this topic is to show how we can borrow from the underlying thinking behind mereology, applying it to transportation systems. There are several advantages to thinking in this manner. To illustrate this point, the next section presents the idea of how to borrow from mereology (very loosely), putting it in the context of transportation systems. This process reveals the advantages of this approach regarding our understanding of many problems associated with urban mobility.

MEREOLOGICALLY INSPIRED THINKING AND TRANSPORTATION SYSTEMS

Essentially, researchers who specialize in mereology are concerned with studying parts and their relations and how parts relate to wholes.[15] In the field of analytic metaphysics, for instance, philosophers examine and discuss these affairs, but it can also lend itself to other matters that would benefit from analyzing parts and part-to-part and part-to-whole relations.

Due to the reality that transportation systems require numerous kinds of parts to operate, researchers and professionals can gain from borrowing the basic structure of mereologically inspired thinking, applying it to their undertakings. Several other scholars have engaged in this practice in many areas of study. Consider, for instance, that work from Frederique de Vignemont et al. demonstrate how to employ mereological thinking to understand how people experience parts of their bodies versus how they experience the whole body.[16] To gain an enhanced perspective of how mereology can benefit engineering, Peter Simons thoroughly explores such benefits in that realm of inquiry.[17]

These previous uses indicate that we can employ mereologically informed insights for guidance, engaging in such enterprises to help develop a way for thinking about transportation on solid ground. It needs to be one that can help create a well-ordered reasoning to support subsequent ontological claims and guide value-based assessments. For these reasons, I am going with unrestricted mereological composition.[18,19] This view maintains that anything can qualify as a part of a transportation system. Though one could make a case for rigorously defining the necessary and sufficient conditions for conceptualizing a transportation system's parts, that notion is only for conceptual direction, bearing in mind that any entity, physical or conceptual, can qualify as a part.

Considering that I am denying any restrictions on a transportation system's composition, I will not offer the necessary and sufficient conditions for parts, relations, or whole transportation systems. Instead, I will examine parts, part-to-part relations, part-to-whole relations, and whole transportation systems. For such purposes, some parts are stand-alone parts, while also being parts of greater wholes, which are also parts of transportation systems. We can break this inventory down into many categories of non-overlapping or disjoint parts, which means that we cannot count a part more than once when accounting for its place in a transportation system.[20] Consider the following example. Trains, tracks, train policies, and train platforms are parts themselves. They help to compose a transit line, and that transit line is part of a transportation system. Due to this structure, the manner that I illustrate transportation systems include strict and metaphoric parts.[21]

When it comes to the parts of a transportation system within a given area, anything that helps people move about the city qualifies as a part. The control of such parts can differ. For example, state or federal entities can maintain authority over some roadways while private owners can operate the vehicles that travel on them. Municipalities or regional cooperatives can direct mass transport services. Cities can provide sidewalks. Private companies can leave scooters on them and college campuses. Bearing the above description in mind, individuals such as train operators can be parts of such systems, which demand that they receive special attention so that they are not treated as just another part.

The motivations behind these parts can vary greatly. For instance, a city can provide walkways and transport services for people to move about the city, help people get to work, engage in commerce, or perhaps an enlightened reason would be to help them thrive. In some cities, municipal entities or private companies can direct rail or bus services concurrently. A state or federal agency could have routed a highspeed roadway or train near or into a metropolitan environment with the primary motivation to promote the distribution of goods, helping industries with supply-chain management. Micromobility start-ups can leave cars, bicycles, and scooters in popular areas with the hopes of making a buck.

Although the above list is non-exhaustive, it indicates the kind of parts that one can expect to find when taking inventory of an urban environment. One can reasonably expect that the diversity of such parts carries over into the unique issues that can arise in different places. Some cities have numerous parts, having to manage their interplay as best as possible. In contrast, other municipalities or regions can have very few parts, ones that are primarily controlled by individual motorists. This notion suggests that we can also anticipate that cities will also have problems of different scales and kinds. That is, there will be problems that affect all users of a transport system. These are macro-level concerns.[22]

They require attention because they largely shape transportation systems in a way that will impact the users who are enduring mobility hardships, as mentioned earlier. These issues also affect all other stakeholders, human and nonhuman, and future people. Due to having such an enormous impact on all parties who have a stake or the possibility of having a stake, macro-level problems require our initial attention and action. Discovering mitigatory efforts, systematic changes, rearrangements, and/or distributions that can remedy mobility ills could reduce or eliminate the need to thoroughly address some affairs.

To familiarize ourselves with these kinds of issues, here is one point that is well known and despised by numerous transportation researchers and professionals: thinking that adding another lane on a highway can solve the problem of having too many cars on the highway.[23] In mereologically inspired terms, it concerns a concentrated arrangement of parts that presents us with a particular conundrum that requires investigation.

This concern is prevalent in numerous cities found throughout the Americas and Asia, "mono-technical saturation." This condition describes the outcome of having an extreme surplus of one kind of transportation part that creates unwanted events while lacking a sufficient number of alternative parts that would ease the burden that such a situation helps create. One could argue that having too many automobiles on gridlocked roadways while lacking effective alternatives adequately accounts for such a phenomenon.

The second condition that I will flesh out in mereologically inspired terms is "poly-technical dispersion," which is a condition wherein several modes of mobility account for the available kinds of parts that can be part of the larger transportation system. One could argue that it is the "antidote" for the ill above. While these terms merely encapsulate conditions known to professionals and anyone who thinks about why such situations exist, as conceptual devices that can facilitate the kind of discussions that are beneficial for mind-frames centered on mitigating the harmful effects of such arrangements. They align in backward-looking and forward-looking fashions.

For instance, with regard to the latter, many contemporary professionals and transportation researchers who design and redesign urban mobility networks tacitly or unknowingly advocate for such an arrangement of parts in their technical vocabulary of multi-modal transportation. While adopting their specialized terms seems as if it were a natural course for argumentation, it lacks the required philosophical longevity to carry out the present task from the ontological to the ethical, which is an encompassing philosophy of urban mobility. The reason why I am sticking with the mereologically inspired approach is that it cannot only help us take note of the transportation system's structure, but it will also lend itself to ethical, social, and political discussions in subsequent chapters.

To address this issue, the following section examines it in some detail. Although the insights into our mobility ills that we can gather from examining this macro-level situation will remain incredibly modest, the point here is to understand how we can employ mereologically inspired thinking to deal with transportation systems in a way that yields insights. In turn, while the example below outlines the characteristics associated with macro-level mobility problems, addressing them properly can also help create better conditions for urban transportation.

URBAN MOBILITY AND MONO-TECHNICAL SATURATION

The setup for this section is somewhat rhetorical. It shapes concepts so that they are compatible with mereologically inspired phrasing to provide theoretical consistency in the following chapters. This point aside, the pages that follow illustrate the conditions that remain connected to some of the pressing issues associated with part-to-part relationships as they pertain to transportation. That is, while most mid-size and large cities and metropolitan transportation systems facilitate mass transit parts such as trains, light rail, and buses, along with bike lanes and wide sidewalks, there are still numerous areas that lack these components as parts of their transportation system. Still, other elements of existing mass transport systems could deter or seriously challenge riders or would-be users. These considerations could include but are not limited to aspects such as ease of accessibility, cost, general feasibility due to constraints such as time spent during travel, and personal safety.

Exploring this issue requires that we expand the scope of our inquiry to include how external parts affect transportation systems. The force of some such elements often shapes the conditions for urban transport systems, which include aspects such as single-family zoning ordinances and parking space requirements for apartments and businesses, which are common in the United States. These factors' influence carries an enormous force that holds steady as conditions that require urban transport plans to work around them.

Some cities and countries in Europe are making strides to be less automobile-centric.[24] However, one could argue that, globally, it will take significant time and attention to secure similar efforts. Overcoming the status quo's resistance to restructuring transportation systems so that they do not favor mono-technical saturation, dethroning automobiles as the top contender in the fight for urban mobility, will not be easy. For cities and regions with mass transit parts that operate alongside a deeply entrenched proclivity toward vehicles and expansive roadways as ubiquitous parts with far-reaching effects, the integrity and true optimization of multi-modal transit will never have an opportunity to display how it can improve urban mobility for all.

This point shows that mono-technical saturation of automobile-centric parts is not only a condition of having too many parts of the same kind. It extends into other parts' ability to have a meaningful place within transportation systems to operate optimally. Instead of such parts being able to operate in a manner that supports other parts that support multi-modal mobility, they must often be situated to support mono-technical saturation, if not in the urban core, then in the suburban areas. For instance, massive "park-and-ride" parking lots where drivers can leave their vehicles to commute by train support this view. Their existence suggests that efforts to reduce the number of vehicles on roadways do not involve decreasing the number of automobiles on the roads—only to lessen the number that is operating during peak hours. In turn, such mass transit parts are only meant to bolster mono-technical saturation's longevity.

We should care deeply about this point because these issues largely shape the conditions of people's existence. Consider, for instance, that instead of having efficient modes of public transport, the primary mode of mobility requires individuals to acquire and maintain personal vehicles, operating them on publicly owned roadways, along with publicly or privately managed roads that require users to pay a fee. As individual "operators" —drivers, are each controlling a private part, a vehicle, in a "collective" manner. Although drivers receive similar training in many instances and are often required to pass the same driving exams, drivers bring their learned behavior patterns, proclivities, habits, and attitudes with them when they are behind the wheel. Along with travel to work, recreation, and errands, some roadway users are doing so because it is their profession, including long-haul truck drivers, transportation network company drivers, taxi drivers, and delivery persons. One could argue that such scenarios are nothing more than slightly organized absurdity. Due to these kinds of arrangements, the number of vehicles as parts of a transportation system far outnumber other kinds of parts within it. These circumstances have led to creating several specialized terms that express the drivers' experiences with these parts. Each one serves as a conceptual shorthand for a situation that would often not make much sense outside of such a context.

For example, due to having numerous drivers who are acting individually but also in concert to make the transportation system "function," frequently it is the case that they fail to communicate appropriately or act in a predictable manner that shows courtesy. This condition can result in "road rage," which is identifiable by verbal expressions of anger and hand gestures.[25] In some instances, such occurrences can lead to fisticuffs, the discharge of weapons, bodily assaults, and deaths. Often, judicial charges are involved that specifically account for such altercations.[26]

While mentioning that this term is rhetorical, doing so would gesture toward the idea that there is an entire vocabulary premised on the notion that urban mobility must include automobiles and their associated roadways as essential parts of transportation systems. Other such terms also serve significant roles in helping us see the depths that accompany the mind-frames that are bound to prioritized automobility-centric urban centers, which are specific instances of mono-technical saturation.

For instance, another familiar term is gridlock, signifying that vehicles have stopped moving or accelerating expeditiously in any meaningful sense while on a major roadway or highway. For such cases, aside from the cause, we can say that such events are instances that stem from mono-technical saturation. Within the mainstream transportation literature, dealing with variations of such scenarios hold steady as areas of focus, which normalizes the view that much attention is being given to making impossible situations more bearable.[27]

Although the examples above focus on automobiles as the primary part wherein there is an overabundance, these kinds of parts are not required for mono-technical saturation. Using personal vehicles as the example for this condition is not meant to vilify them. There is nothing inherently wrong or bad about cars. Despite the reality that this situation happens in numerous places only speaks to the prevalence of the problem. The same situation could exist in regions where overcrowded train cars are the only parts in operation. Similarly, imagine a scenario involving a transportation system that only used bicycles, a situation that could give way to bicycle traffic jams.[28]

While there could be numerous examples that show how having an overabundance of the same kind of part could lead to mono-technical saturation, the idea to keep in mind is that this situation makes the experience of urban mobility one that people often dread. Consider, for instance, that people often refer to having to "fight" traffic. The use of this term signifies that engaging in the operation of motor vehicles at peak times requires urban dwellers to experience a stressful activity that has become routinized. In turn, people pursue numerous activities while driving to alleviate the unfavorable conditions associated with commuting via automobile when these parts are the primary mode of mobility.[29] Some research shows that while some commuters do other things while driving, they largely consider it as a waste of time.[30] Due to such situations, there is a bounty of public health research that studies the ill effects that we can argue stem from mono-technical saturation.[31]

Although the above conditions are unfavorable, despite being a commonplace occurrence, one could speculate that numerous people passively accept them as a permanent reality, unaware that another world of urban mobility is possible. Through providing alternatives, such changes could relieve the stress that many people experience, arguably improving their lives. During the outset of this chapter, I mentioned that we could change these conditions, and the study of part-to-part and part-to-whole relationships can help create better outcomes. In turn, the section below goes in that direction.

URBAN MOBILITY AND POLY-TECHNICAL DISPERSION

One primary challenge to addressing mono-technical saturation is that the infrastructure that supports it, namely roadways and bridges, is difficult and expensive to remove or reconfigure, and such drastic measures are arguably controversial. Still, highway removal projects often benefit residents.[32] Although precedent cases are rare, those available suggest that undertaking such tasks are welcomed changes to existing transportation systems. Yet, the task is to increase the variety of mobility options so that alternatives remain feasible. Unless this condition is met, one can argue that such an attempt to go against mono-technical saturation will be too shallow to count toward substantial progress.

According to several transportation scholars and professionals, the remedy for the ills associated with what I have described as mono-technical saturation is to amend cities with multi-modal transportation systems.[33] The basic idea is that instead of having a dominant form of mobility, such as personal vehicles, urban spaces should include many other forms such as light rail, buses, bicycles, and wide sidewalks. The motivation here is to provide alternative modes of mobility that people can readily access, a necessary step to put the car keys down and explore other ways to move about the city.

Putting such insights into an applied mereological context suggests that we need to increase the kinds of parts that help compose a transport system as a whole entity. Achieving this process, either as a mitigatory effort or as a feature of the design, counts as poly-technical dispersion. That is to say, making several kinds of parts available as feasible alternatives, accessible city or region-wide, describes the type of response that is necessary to dismantle the harmful and/or unwanted conditions described above. Consider, for example, if the primary means of urban mobility were personal automobiles and roadways, which led to conditions such as daily gridlock, then the challenge is to increase the options for people.

Bearing this point in mind, mono-technical saturation is an expression of the tension within many transportation systems across the globe. In turn, poly-technical dispersion describes the manner wherein professionals can work toward easing it. Making room for additional means of mobility serves as a valve, planning efforts that make urban mobility less stressful and, perhaps, enjoyable. Considering the wide variety of issues that, when examined, expose the numerous harms that stem from transportation systems, the idea that they could be a source of joy might sound far-fetched. However, this is not the case for many travelers. Several people who commute by bicycle report high degrees of satisfaction.[34] The city of Portland, Oregon, was not always a paradise for bicyclists, but it came to be that way through community-engaged policy initiated in 1973.[35] Today, the city has a reputation for its ubiquitous bike lanes, attracting residents who want to pedal rather than drive.

Additionally, some people prefer to leave the driving to professionals, opting for public transport options such as buses and light rail for reasons other than necessity or efficiency. These riders fall into categories such as "occasional," "commuter," and "all-purpose" riders.[36] What is more, some bus riders report feeling that they gain a sense of community from riding the same bus line.[37] Many such people often hold the bus driver in high regard, thanking him or her for the lift.

The point in mentioning the two cases above is not to fetishize modes of transit. Rather, it illustrates that mobility options inherently increase the quality of people's lives while other modes can go in the opposite direction. These points indicate that if mobility options—when thought about as automobile-centric planning as "the only game in town"—support mono-technical saturation, then venturing outside this pattern of thinking can lend itself to developing a poly-technical dispersion of means to secure urban mobility. One can push back against this point, holding that people engage in practices while driving that they enjoy. For instance, people listen to music, podcast, and books while commuting or moving about the city. Such activities can obviously enrich their lives. In turn, by urban travelers having the ability to pursue these pleasures while driving is basically the same as any other way to navigate the cityscape. What is more, some people enjoy driving.

This rebuttal is significant. However, there is a primary difference. For the former, the transit choice is inherently pleasurable. In the latter case, those activities are ones that the driver can pursue to make the drive more enjoyable—in certain instances. If they were not put in a position so that they *had* to operate a vehicle, then they would not have had to find an activity that made the drive more bearable or enjoyable. Despite this condition, it could be the case that they found both driving and listing to entertainment to be equal to or greater than being a passenger on a bus or having the option to ride a bicycle.

Fair enough. I am willing to concede such a possibility exists. However, it does not take away from the reality that the other possibility exists wherein the preferred mode of mobility could very well be a more liberating choice for the user, seeing as how they have the option to engage in the mode of transit. Still, further, advocating for the automobile-centric status quo does nothing to encourage poly-technical dispersion as a means to reduce or eliminate mono-technical saturation. This situation is the one that has created the numerous problems mentioned during the outset of this chapter that still require attention.

Before we can begin to address how to deal with mono-technical saturation, we must address a range of additional concerns. Once revealed, they will help expose the totality of transportation's effects that must come into view to gain a comprehensive account of urban mobility. Such an undertaking will also contribute to laying down the moral "suggestions" for how we go about developing mitigatory efforts to the many issues mentioned previously. For instance, one of the vital points that I make is that we first need to talk about the prioritization of the stakeholders who will be impacted by such a conversation. Through dealing with this point, we can continue to answer the question, "What is a transportation system?"

This move is necessary because the manner wherein we decide who should receive consideration, followed by action, is subject to debate, and attending to the latter dimension of this inquiry reveals the impetus behind such discussions. Chapter 4 begins to address these concerns. This theme continues throughout the next few chapters, providing insights into the necessary elements that must align to see how this approach's trajectory offers advantageous reasons to support thinking about urban mobility as a complicated affair that deserves this kind of attention.

NOTES

1. For a "textbook" account of urban mobility, Marcus Foth defines it as follows: "Urban mobility: The whole of trips generated daily by the inhabitants of a city, and the methods and conditions associated with such trips (modes of transport selected, length of trip, time spent in transport, etc.)" For more information, see Marcus Foth, *Handbook of research on urban informatics: The practice and promise of the real-time city* (Hershey, PA: Information Science Reference, 2009), 389.
2. For an example, see *Southern Nevada Strong*. http://sns.rtcsnv.com.
3. For an example, Briana Payne describes how the Bonton and Ideal neighborhoods in Dallas, Texas have this kind of history. For more information, see Briana Payne, "Oral history of Bonton and ideal neighborhoods in Dallas, Texas," MA Thesis University of North Texas, 2015, 127.

4. Karel Martens, *Transport justice: Designing fair transportation systems* (New York City: Routledge, 2016), 22.

5. Shane Epting, "The Peñalosa Principle of Transportation Democracy: lessons from Bogotá on the morality of urban mobility." *Science and Engineering Ethics* 23, no. 4 (2017): 1085–1096.

6. Erin Cornwell *et al.*, "Social disconnectedness, perceived isolation, and health among older adults," *Journal of Health and Social Behavior* 50, no. 1 (2009): 33.

7. Vera Tillmann *et al.*, "Public bus drivers and social inclusion: Evaluation of their knowledge and attitudes toward people with intellectual disabilities," *Journal of Policy and Practice in Intellectual Disabilities* 10, no. 4 (2013): 307.

8. For information on transgender and gender non-conforming persons and ill-treatment on public transport, see Amy Amy Lubitow *et al.*, "Transmobilities: Mobility, harassment, and violence experienced by transgender and gender nonconforming public transit riders in Portland, Oregon," *Gender, Place & Culture* 24, no. 10 (2017): 1398. For a specific case the examines women and mobility services, see Sana Iqbal, "Mobility justice, phenomenology and gender: A case from Karachi," *Essays in Philosophy* 20, no. 2 (2019): 171.

9. Robert Schneider *et al.*, "Comparison of US metropolitan region pedestrian and bicyclist fatality rates," *Accident Analysis & Prevention* 106 (2017): 82.

10. Evelyn Blumenberg *et al.*, "Getting around when you're just getting by: Transportation survival strategies of the poor," *Journal of Poverty* 18, no. 4 (2014): 356.

11. Juliana Maantay, "Asthma and air pollution in the Bronx: Methodological and data considerations in using GIS for environmental justice and health research," *Health & Place* 13, no. 1 (2007): 40.

12. Briana Payne, "Oral history," 127.

13. Aaron Scott, "By the grace of god," *Portland Monthly*, March 2012. https://www.pdxmonthly.com/articles/2012/2/17/african-american-churches-north-portland-march-2012.

14. Trina Rytwinski *et al.*, "How effective is road mitigation at reducing road-kill? A meta-analysis," *PLoS One* 11, no. 11 (2016): e0166941.

15. Achille Varzi, "Mereology," in *The Stanford encyclopedia of philosophy*, ed. Zalta, E.N. 2015. http://plato.stanford.edu/archives/spr2015/entries/mereology/.

16. Frederique de Vignemont *et al.*, "Body mereology," in *Human body perception from the inside out*, eds. Gunther Knoblich, Ian Thornton, Marc Grosjean and Maggie Shiffrar (New York, NY: Oxford University Press), 148.

17. Peter Simons, "Varieties of parthood: Ontology learns from engineering," in *Philosophy and engineering: Reflections on practice, principles and process*, eds. Diane Michelfelder, Natasha McCarthy and David Goldberg (Dordrecht: Springer, 2013), 152.

18. While there are other issues with unrestricted mereological composition, I am only applying it to the context of urban mobility in this chapter.

19. For more information on this topic, see Timothy McCarthy, "A note on unrestricted composition," *Thought: A Journal of Philosophy* 4 no. 3 (2015): 202–11.

20. For a good discussion on the specific topic of non-overlapping parts, see Giorgio Lando, *Mereology: A philosophical introduction* (London and New York City: Bloomsbury Publishing, 2017), 61.

21. As Giorgio Lando argues, differentiating between literal and metaphorical parts can be exceedingly difficult at times. To account for such a condition, I steer away from these debates. Bearing this point in mind, I am sticking to discussion on strict parts unless I state otherwise. For more information, see Giorgio Lando, *Mereology: A philosophical introduction* (Bloomsbury Publishing, 2017), 24.

22. While this chapter examines mono-technical saturation as a macro-level issue, other such issues could exist that have not been identified. Additionally, it could be the case that similar concerns could emerge in the future that would follow the pattern of mono-technical saturation.

23. Ronald Milam *et al.*, "Closing the induced vehicle travel gap between research and practice," *Transportation Research Record* 2653, no. 1 (2017): 10.

24. Mark Nieuwenhuijsen *et al.*, "Car free cities: Pathway to healthy urban living," *Environment International* 94 (2016): 257.

25. Leon James, *Road rage and aggressive driving: Steering clear of highway warfare* (Amherst, NY: Prometheus Books, 2009), 21.

26. James, *Road Rage*, 21.

27. For an example of this situation, see Hani Mahmassani *et al.*, "Urban network gridlock: Theory, characteristics, and dynamics," *Transportation Research Part C: Emerging Technologies* 36 (2013): 480–97.

28. It is my understanding that traffic jams that involve only bicycles are quite common in the Netherlands. For an example, see https://www.youtube.com/watch?v=qp0cFOubQ5I.

29. Patrick Singleton, "How useful is travel-based multitasking? Evidence from commuters in Portland, Oregon," *Transportation Research Record* 2672, no. 50 (2018): 14.

30. Singleton, "How useful," 16.

31. For an example of such effects, see Mehrdad Tajalli *et al.*, "On the relationships between commuting mode choice and public health," *Journal of Transport & Health* 4 (2017): 267–77.

32. City of Seattle, "6 case studies in urban freeway removal," *Seattle urban mobility plan.*

33. For an example of such positions, see Sabyasachee Mishra *et al.*, "Performance indicators for public transit connectivity in multi-modal transportation networks," *Transportation Research Part A: Policy and Practice* 46, no. 7 (2012): 1066.

34. Kirsty Wild *et al.*, "Why are cyclists the happiest commuters? Health, pleasure and the e-bike," *Journal of Transport & Health* 14 (2019): 100569.

35. City of Portland, Office of Transportation, Bicycle Master Plan. https://www.portlandoregon.gov/transportation/article/369990. Last updated, July 1, 1998. 1.

36. TransitCenter, *Who's onboard 2016*. https://transitcenter.org/wp-content/uploads/2016/07/Whos-On-Board-2016-7_12_2016.pdf, 5.

37. Jeff Nash, "Bus riding: Community on wheels," *Urban Life* 4, no. 1 (1975): 100.

Chapter 4

Moving, Parts, and Morality

While urban transportation helps move people about the city, it also helps them lead flourishing lives. This thrust cuts through many layers that, once revealed, show that the functionality of urban mobility systems only begins with the need to transport people to and from destinations. In addition to these purposes, they also play moral roles in the outcomes they coproduce with other socio-material parts of the urban sphere: all the city is a stage, and the residents are the free-moral actors. They are the most significant parts with the highest kind of intrinsic value. Due to this consideration, they are "active" parts, even though referring to people as parts sounds a bit off.

This point aside, a case could be made that a person would like being thought about as being "part" of his, her, or their city, even though they do not want to be reduced to being thought about as merely *a* "part." While saying that these two senses of the word "part" are the same would be wrong, both suggest that the city needs people to exist, even though they exist as parts. Calling people "parts" might sound weird, but this situation is merely a condition of language. Yet, to qualify this position, we can say that, only in mereologically inspired terms, are people thought about as parts. Outside this context, there is no reason to make such a claim. They are "active" parts because they move about the city, and move, steer, operate, use, and/or ride the other parts, directly and indirectly. Without them, of course, there is no reason for the other parts, which we can label "passive" parts.[1] These are the parts that form the city's backdrop. Together, in the context of urban mobility, active parts make the (passive) trains run on time.

For transportation specialists, they play active *lead* roles in a capacity that shows they remain indispensable. Fulfilling a fundamental operation in the daily lives of urban dwellers exhibits their significance. Providing these essential needs to numerous people not only gives them the required means to navigate cityscapes, but also give people a way to make the most out of the opportunities that city life makes possible within the urban environment.[2] With this notion in mind, this chapter, as an excursion into the next phase of framing elements of transportation systems as parts that fit together to deliver desired outcomes, begins. It explores what it means for mobility systems to be wholes that can help us create socially just and sustainable futures for the city that align with Jonas's imperative.

In the pages and chapters that follow, we will see how transportation is not just vital for urban living, but also how it helps people to adapt to challenging conditions such as climate change, political turmoil, power consolidations, vast economic inequalities, and global health issues. The notion that requires ample underscoring here is that mobility systems must be incredibly adaptive to address ethical obstacles that emerge in the city. Such a condition brings significant issues that go incredibly far beyond contemporary moral affairs such as promise-breaking and thievery.

For instance, cities constantly and rapidly change. This reality entails that the circumstances that accompany these shifts will be far more frequently than other kinds of concerns. With this point in mind, the focus going forward centers on developing a way to deal with issues that arise from a shifting set of conditions, an acutely adaptive approach. Aside from people, the parts of a transportation system are interchangeable. We can move, add, or remove them. This situation opens up significant possibilities. It also signals that we must address these issues in several ways to understand the robust nature of transportation systems and the myriad problems connected to them.

On the meta-level, if the conditions surrounding mobility problems are continually shifting, then the conditions that pertain to creating mitigatory efforts should also do the same. This notion indicates the sort of dynamics that we can expect to encounter when locating the pattern behind transportation systems that could account for all of their relevant parts. Such an idea entails that we go beyond the parts that municipal, state, national, and/or private agencies control. It also includes the auxiliary (active) parts such as drivers and (passive) parts such as home-owners' associations that can play a role in the functioning of urban mobility.

When bringing these aspects into our purview, it is a sobering realization to note that the background for the issues at hand remain incredibly fluid, considering the almost impossible reality that we cannot simply keep track of all such parts. While this idea might seem daunting, instead of pretending that urban mobility problems rest on a solid and stable foundation that leans heavily toward the universal, grounding our thinking to learn and act from a real-world reasoning can deliver better outcomes sooner—instead of never.

The point here is not that searching for absolute principles is not a worthwhile endeavor. It is. However, engaging in a practice that requires dedication to focus on the abstract rather than the dynamic and concrete problems that we can readily tackle says something about our priorities for problem-solving, which is itself a moral affair.[3] These are the numerous concerns listed in chapter 3 that speaks to this point's fundamental integrity. Not wanting to address existing problems until larger abstract matters receive attention could prolong suffering and injustice. This point should force us to recall an aphorism that Dr. Martin Luther King, Jr. employed in his famous *Letter from Birmingham City Jail*: "Justice too long delayed is justice denied."[4]

Drawing from his wisdom, the way that we are accustomed to dealing with real-world issues philosophically involves looking back abstractly on principles that inform us. In turn, we need to balance both kinds of thinking, often but not exclusively yielding to real concerns when conflicts of our attention and priorities arise. Analyzing the problem is one issue, but acting is another. Considering that we are dealing with human lives, I argue that we should yield to the reverend's insights. If mistakes are made and not too grave, then at least there is something to remedy rather than no movement at all. The former shows care while the latter shows nothing. While I cannot speak for anyone else, going with the former at least can provide some degree of comfort and, possibly, hope. Moreover, it suggests that the attitudes accompanying mitigatory efforts are geared toward people's well-being, even though they could require adjustments and/or reconceptualizing in some instances.

By developing an approach with such an orientation and these characteristics, it positions us to work to improve the livelihoods of billions of people who must contend with transportation issues. Prioritizing our efforts in this manner makes a social, political, and moral statement, saying that people's lives and livelihoods matter more than scholarly discovery and debates. If we think about urban mobility in this manner, it can help us restore the conditions for nonhuman life to also thrive, along with the mitigatory efforts that come with such aims.

Additionally, while the conditions that cities must face that concern hardship today will constantly change, transportation specialists must also plan for tomorrow's residents and their mobility and well-being. Such scenarios suggest that change is inherent to cities in two senses: existing in the present and in the future. While it is common to anticipate transportation planning for future needs, the idea of "tomorrow" does not merely signal that we need to plan for immediate urban mobility needs or set some resources aside for the next generation. It goes beyond such a narrow scope. Considering that numerous cities are hundreds and thousands of years old, we can at least conceptualize that such conditions could still apply if global climate change does not render them inhabitable for most people.

Still, setting our sights on such a short timeframe only helps us expand our horizons in the very briefest of senses. This conception would not be a problem if our mobility troubles could be solved entirely and permanently. However, one way to think about the distinct nature of such issues is that their ends get further away as we approach them, a notion that practically evades our imaginative powers. That is, as cities increase or decrease populations, annex additional lands for new neighborhoods and businesses, open stadiums, parks, and recreational spots, transportation must accompany these changes. Thinking about the complex realities wherein an industry's success or failure could create or destroy a municipality's mobility needs suggests that we cannot conceive of transportations systems as static. Although numerous markers signal change in a given city, anticipating such an event remains challenging in the best instance and impossible in the worst case. Moreover, one could argue that many people's relationships exist within a given set of durations, which remain inherently limited to the present or the near future. This point suggests that thinking of the "future" as extremely distant is not realistically fathomable under many circumstances.

Consider, for instance, that to even think about mobility in the future requires that we enhance our conception of time to go beyond our lifetimes or our great, great-grandchildren. This notion entails that we probably do not even know how to begin thinking about what could come 500 years from today. Entertaining this idea means that we need to be cognizant that modes of urban transit that will exist in the world beyond tomorrow escape our limited thinking on the topic. In turn, we are significantly challenged when it comes to envisioning the situated conditions that will surround urban mobility quite distantly, especially when so many other non-transportation related parts will also influence it. We also cannot anticipate those factors, which we could call "external," even though they remain relevant.

Although our best minds cannot determine exactly what the world of urban mobility will look like in the future, the condition that we must reconcile, as Hans Jonas frames it, is that we must not destroy the conditions in the present that will allow people in the future to exist fully and genuinely.[5] Giving this sentiment respect demands that we must continue to create places where people desire to live while developing the ability to move about them reasonably. Keeping this idea in mind, we cannot let the means to this end possibly destroy the possibility of having ends.

This notion implies that cities must continue to pursue worthwhile goals that make people love living in the city, even as we contemplate the definition of what it means to live in one. In turn, extending the pattern of the ideas above demands that we think about the specific characteristics that help give identities to the places that we call home. In particular, we need to examine how transportation systems as wholes can help create a world that we want to materialize. Specifically, there are two lessons from the philosophy-of-technology literature that can help us in this endeavor.

One such advantage is that we can zero in on technological parts in different scales and contexts, learning how they affect and are affected by surrounding parts, along with external social and political systems and ecosystems. The reality that transportation systems are ecologically, socially, and historically situated increases the layers of consideration. Gaining a comprehensive view of these elements can inform the litany of issues that we find in Jonas's concerns about modern technology's ethical dimensions. These topics are realities that require immediate attention in some regions, and it is reasonable to hold that other locations will encounter similar situations that arise from the enmeshment of urbanization and climate change.

For instance, urban sprawl, desertification, and wildlife displacement are exemplars of environmental cases. Bearing in mind that many people long forget events such as destroying neighborhoods that once belonged to marginalized groups for highways, these incidents exhibit that such cases have preexisting conditions that require consideration alongside new and emerging problems such as those just mentioned. Should we not bring them into our purview when moving forward? If we should, then we must include these kinds of considerations when contemplating how to advance our thinking regarding such matters. Undertaking such a task is no easy feat. Highly complex issues such as those above could utilize lessons from highly advanced research in philosophy, bearing in mind that moral progress should accompany technological advancements.

Studying the progress in the philosophy of technology can also provide a starting point to identify the range of harmful and positive technological impacts that the arrangement of parts can have on several groups. Such conditions can manifest in numerous ways, affecting stakeholders such as vulnerable and/or marginalized populations, the public, nonhuman life, future humans, and anthropogenic urban artifacts. The dimensions of such effects that are immediately noticeable include air quality, water safety, resource scarcity, and environmental pollution.

Yet, due to the enmeshment of mobility networks with other urban artifacts such as homes, office buildings, commercial districts, industrial centers, and places for recreation, dining, and entertainment, isolating transportation services to become aware of the associations between them and the results of long-term environmental degradation remains challenging. This notion suggests that it might be too difficult to identify some kind of inherent moral quality that belongs to a particular transportation technology, meaning that we have to study them as they appear in *present* urban contexts, which includes their cumulations and histories.

Even though these latter dimensions are not immediately present to all people, the felt effects are significant, and some groups have made this point abundantly clear.[6] This notion underscores the idea that intentionality is mostly irrelevant when stacked against the outcomes that determine the quality of people's lives, from the communities that were destroyed for additional roadways to the people who must contend with pollution. The reality is that one single part of a transportation system cannot play such a significant role without an arrangement of parts that creates the conditions wherein harmful outcomes emerge. Mono-technical saturation of automobiles would not be possible if it were not for having plenty of roadways, a lack of adequate alternatives, and any other elements that play "roles" in scenarios that help create bad outcomes.[7]

For instance, if one could travel back in time, she would be incredibly hard-pressed to convince Karl Benz that his gasoline-powered automobile would play a role in changing the planet's climate, along with shaping cities and billions of people's lives. This idea suggests that vehicles are not inherently harmful parts, but something there requires additional investigation. That is, numerous times over, gasoline-driven vehicles are linked to ongoing events such as climate change. Even though they are not entirely responsible for all of the ill effects of a warming planet, convincing the scientists behind this research that there is no connection between auto emissions and climate change would be almost impossible, especially considering that their views rest on numerous sound studies.

Taking this point seriously means that when weighing gasoline-powered (personal) vehicles against other modes of transport, all of the effects developed over more than a century must be brought into view, which includes numerous aspects beyond climate change. For instance, having automobiles as the dominant mode of mobility means that neighborhoods lack planned walkability, affecting public health.[8] The impacts of mono-technical saturation go far beyond gridlock, as examined in chapter 3, resulting in a range of felt effects on people.

For starters, populations living near highways often suffer disproportionate rates of respiratory illness.[9] Another issue is that there are economic burdens associated with automobile ownership. Select instances show that many people are "forced" to purchase private vehicles to accomplish the tasks related to basic living.[10] In turn, when, on a global scale, the ebb and flow of oil prices hit consumers, many of them lack a realistic alternative to privately owned vehicles or efficient mass transport. These are just a few of the kinds of concerns that require attention. Yet, they indicate the types of issues that occur often and compound over time. In turn, there is a pressing need to examine how transportation affects human and nonhuman worlds. To undertake such a task, there is a need to conceptually map all of the parts, necessary considerations, and exceptions.

For such a task, examining popular positions in philosophical subfields such as philosophy of technology and environmental ethics can guide us, with some limits. While additional insights from these fields can benefit how we lay out the areas of concern that require further inquiry, we must confront a few areas within specific scholarly debates to advance the efficiency and congruency needed to think through urban mobility. Although ironing out some of the conceptual wrinkles in the fabric of intra-disciplinary thought might initially appear as a monumental move that impedes the ability to address concrete transportation issues, attending to such matters will solidify the grounding that is required to build a foundation that can help secure those outcomes.

While pursuing this kind of enterprise at this level takes us into mostly uncharted territory, getting it right might require that we make some mistakes along the way. This notion underscores the idea of why exceptions must hold steady as an intricate dimension that belongs to novel ways of approaching complicated issues such as those that we encounter when bringing the full range of stakeholders into view. It provides the opportunity to make adjustments on a case-by-case basis, along with the means to remedy troubling aspects that yield shortcomings, theoretical or concrete. The following section moves us in that direction to explore and flesh out these affairs, paying attention to the inevitable complexities that remain inherent to complicated transportation issues.

STRUCTURAL ETHICS AND TRANSPORTATION TECHNOLOGIES: TOWARD MORAL ORDERING

Having situated our thinking that transportation systems are whole technologies that have numerous smaller parts provide many benefits. The primary advantage of underscoring the notion that transportation is a technology is that wrapping our heads around its effects becomes an intelligible and manageable task. To achieve this end, recent advancements in the philosophy-of-technology literature give us an excellent backdrop to see how mobility technologies help shape morality in the city. Specifically, the structural ethics approach has recently emerged, showing promise to help us deal with relationships involving humans and technologies with moral dimensions.[11] In turn, one can make a case that this approach is indebted to many proceeding views holding that technologies have agency, either equal to human agency or some degree thereof.

For example, there are two extremely well-known cases. The popularity of such positions has roots in Bruno Latour's work, evident in his view that technologies are the missing masses, a network with humans wherein they frequently stand in for us.[12] Latour uses a memorable example to illustrate his point. That is, he humorously holds that his automatic seatbelt, through a cabal of clever engineers and authorities, makes him behave morally. He cannot escape wearing a seatbelt—unless he solicits the assistance of an unscrupulous mechanic.[13] That point aside, Latour's position is tantamount to the view that machines can act in a sense, serving as the missing moral masses in society. We often talk of morality being absent from the public sphere, but it is there in the form of devices that stand in for us.[14]

After Latour, alternative positions in the philosophy of technology emerged that offered a less intense view of nonhuman agency.[15] Perhaps that most well-known argument that puts some distance between ideas such as Latour's that does not strive for full-blown agency is Peter-Paul Verbeek's approach known as *technological mediation*.[16] At the heart of his approach rests the idea that, similar to but separate from Latour, agency remains a property of the networks of humans and devices (nonhumans), underscoring the substantial influence that technologies have on our agency. In turn, they "mediate" our mental and physical life, helping shape our existence and decisions. Such a view shows our agency as essentially compromised, meaning that it is better to describe it as something that works in concert with devices. Here is how Verbeek puts it:

> Moral agency should not be seen as an exclusively human property; it is distributed among human beings and nonhuman entities. Moral action is a *practice* in which humans and nonhumans are integrally connected, generate moral questions, and help to answer them. . . . Everyone, for instance, who slows down near a school because there is a speed bump on the road shows steered behavior rather than moral and responsible action. . . . Here technology does not *impede* morality, but rather *constitutes* it.[17]

While the passage above encapsulates Verbeek's view, along with how he puts a degree of modest distance between himself and Latour, his position remains a close neighbor. Although Latour does not explicitly employ the language of agency in the work on the missing masses, it nevertheless entails it, bearing in mind the claims that he makes wherein humans and nonhumans remain interchangeable. Still, Verbeek does not cast aside appeals to or the need for an account of nonhuman agency, at least in some capacity, clinging to it in a form that he sees as inherently different from Latour. For instance, Verbeek succinctly makes this point evident when he says, "The position that I have laid out . . . is based on the idea that the moral significance of technology is to be found not in some form of independent agency but in the technological *mediation* of actions and decisions—which needs to be seen as a form of agency itself."[18]

It is easy to appreciate his view because, similar to Latour, it spotlights the significant ways that technologies can define our lives through their constant presence. His works in this area have served the philosophy of technology, as a subfield of philosophical (and interdisciplinary) research, incredibly well when exploring the intricate matters that we associate with the myriad complex relationships between humans and technologies. Scholars continue to employ this way of thinking to deal with similar problems today, showing how technological agency maintains its utility for such discussions.

Despite the usefulness that his approach provides, there are at least two reasons why we should continue to explore other ways to provide insights into human-technology relations. The first concern comes from the inherent problems that pertain to his position's structure. Consider, for instance, that Philip Brey puts forth significant challenges for technological mediation in the form mentioned above. For example, Brey argues that:

> While I agree with Verbeek that human agency is often influenced by artifacts, and I am even willing to agree that agency can be attributed to human-artifact assemblies, it does not follow that artifacts therefore have some form of agency, as Verbeek sometimes claims. This is like saying that because salty water is liquid, and it includes salt, that therefore salt has properties of a liquid. It would seem more correct to say that the salt in salty water mediates or transforms the liquidity of water without having liquid properties itself.[19]

Through examining Brey's critique of Verbeek, one can mine it further to gain insights into the exact nature of the problem that troubles the agency of technology. That is, Brey is essentially saying that Verbeek's reasoning is guilty of claiming that while we can say that a human-technology relation involves agency, mediated or not, it does not follow that the nonhuman "part" of that relation also has agency, a fallacy of composition. He assumes that a part has the qualities of the whole, and that notion, one could argue, remains inherently challenging to prove or make a convincing case with firm support. Continuing a line of thought in this direction, Brey argues that:

> Verbeek has not demonstrated that human agency cannot exist independently from artifacts, and that therefore human beings cannot be conceived of as (moral) agents independently of the artifacts they use. Surely, it would seem, human beings that are bereft of any artifacts can still deliberate, intend or act. Humans are moral agents that continually couple with and decouple from artifacts that co-constitute their agency. Verbeek's view therefore gives too much credit to artifacts in assigning agency to them and too little to humans in denying them agency independent of, and prior to, any artifacts they may use.[20]

For this reason alone, one could make a case that we need to rescue the concept of technological mediation from positions that include discussions of agency. The goal here would be to elucidate how technologies impact humankind while keeping human-driven and grounded responsibility firmly in view. Brey's work, as examined a bit below, does just that. However, Brey and Verbeek's irreconcilability holds very little importance for the other reason why agency and elaborate human-technology relations should be of little interest outside of such esoteric debates. In the context of an outcome-driven examination, it is in danger of being more of a red herring than it contributes to how we understand the outcomes of such relations.

Putting the matter into more concrete terms: when dealing with transportation affairs that involve vast physical, political, social, and environmental elements, topics such as agency and intentionality are peripheral issues for scholars who are dedicated to those subjects rather than people who aim to think through mobility issues to deliver "better" situations for society. The outcomes are the focus.

This point is not meant to discount—in any shape, form, or fashion—the bounty of research from urban thinkers and historians who have strong cases that expose racism in transportation planning and engineering. Instead, gesturing toward widely applicable principles that zero in on the outcomes are what "matter," defining this term in a way that shows respect for the people who are harmed or who could be harmed due to such arrangements. Receiving such treatment occurs because it is an inherent product of the system. Technological arrangements in socio-material systems such as transportation can easily lead to good and bad outcomes, intentions and motivations aside.

When we are concerned with producing better outcomes, there is simply no need to bring the alleged "agency" of technologies into question to understand how technologies in our lives and affairs impact people. That is, we simply do not need to discuss it, and giving it serious consideration would take us beyond the necessary bounds of an outcome-focused inquiry. The best way to think about it—outside of the leading research strands in the philosophy of technology—is that the question of technological agency is a superfluous aspect that does not help us advance the human-technology conversation as described above.

The significant challenge to such a view is that conversations about "rigor" tend to dominate in mainstream philosophy and academia. That is, one could push back against the position above, holding that examining technological agency is necessary to have a rigorous investigation of the interplay between humans and devices. Yet, how much rigor is too much or too little for these kinds of studies? There are at least two possible answers. One, if we need to find the "golden mean" of rigor, would it not make sense to say that developing an approach that seriously weighed the significant elements for consideration should demand the spotlight instead of peripheral concerns? This notion suggests that we need to be rigorous when analyzing outcomes and the arrangement of (mobility) technologies that play a role in their creation. Alternatively, suppose we adhere to the proselytization of rigor without limits. In that case, we could say that to give topics the attention they deserve, separating the "agency" of transportation technology from its associated outcomes could help this endeavor. The motivation behind this attitude is that we could advance both areas of study by not having researchers pay attention to subjects that might not be of immediate concern to them. Aside from researchers' penchants for highly specialized topics, other practical considerations warrant additional thinking behind this point.

For instance, when it comes to staying on the task of dealing exclusively with outcomes, I champion the view that gets rid of the alleged "agency" of technology when it comes to approaches that are developed to address ethical situations in existing and neglected technologies such as transportation systems (at least by philosophers, as Karel Martens argues).[21] Recall that during chapter 1, I referred to the idea that we need to remove or "chop" parts of theories that only slowed us down in our thinking—in a manner that does not offer any significant advantages for understanding the multifaceted dimensions of transportation affairs. For the case at hand, technological agency aligns with the definition of a theoretical aspect that needs to be chopped.

Recalling a point from the introductory chapter: the difference between a traditional motorcycle that most people ride and a customized chopper rests on two principal ideas. First, a regular motorcycle was built with the average motorcycle rider in mind. This kind of technology has several parts that make it operate smoothly and efficiently, providing an enjoyable ride. Such motorcycles were designed for just about any person who wished to own one. Contra to such broad appeal, a chopper markedly differs. The non-essential parts have been "chopped," removed so that it can travel at faster speeds safely, and it has been customized for enhanced performance.[22] To illustrate that this bike is associated with only a specific rider, it has been personalized aesthetically.

In a similar fashion to motorcycles, customization for dealing with transportation issues in real cities must hold steady as a central tenet to our thinking about such matters due to the unique characteristics that belong to them. Due to such aspects, this reason is precisely why we need a flexible, adaptable "anti-framework" framework to make moral sense of transportation issues. To approach any such affair in urban mobility will require a common moral language. However, it would not benefit from a rigid moral framework that expects the world to stand still.

Pursuing such goals would benefit significantly from how structural ethics, as a system that analyzes humans and technology's interplay, focuses on the outcomes that such arrangements produce. This approach's primary advantage is that it positions technologies as "playing" or "having" moral roles in societies, viewed as elements within multifaceted arrangements of other social and physical systems.[23] This approach does not view technology as being entirely and inherently positive or negative. However, we can employ those terms to the outcomes that they help produce within those larger socio-material structures. This characteristic of structural ethics applied to the present case means that we think about transportation technologies in a way the zeroes in on the idea that we can direct them in broader varying sociopolitical and ecological surroundings, working toward the results that we favor.

Putting structural ethics in terms that are compatible with applied mereology, transportation parts become "moral parts" when situated in contexts that support the desired (moral) outcomes that improve urban mobility in a given capacity.[24] By engaging in this kind of thinking, we bring our responsibility for technology within human-technology relationships into view, paying attention to the choices we can make to use technology to deliver the results we want. In turn, this approach gives a way to guide the implementation of particular technologies, remove harmful ones, and work toward human-technology relationships that foster worthwhile goals such as socially just sustainability and human flourishing.

Suppose we apply this approach to transportation planning and engineering. In that case, it can guide its development, improving the likelihood that we can achieve the goals that we advance because it offers guidance. Despite this strength, structural ethics must expand its coverage to guide outcomes beyond basic human-technology relationships, primarily in the following ways. Specifically, it needs to take inventory of how a given technology will affect a range of stakeholders. As mentioned previously, they belong to several categories, including vulnerable and marginalized people, the public, nonhumans, including individual species and ecosystems, future generations of people, and urban artifacts such as buildings, bridges, and ballparks.[25]

Expanding structural ethics in this way means that there is a need to discuss how the effects on specific categories will carry more moral weight than other groups. This notion bears significant importance, especially considering that we do not want to give the same degree of consideration to vulnerable people that we do to bridges. For transportation issues, major infrastructure and policy projects will significantly impact all of the groupings above. Weighing such interests, especially stacking nonhuman interests against those of humans who presently exist, is a highly intricate affair.

Despite having to deal with such complicated issues, there is a history in philosophy that we can trace, from existentialism to environmental ethics, that exhibits how progress in philosophic research can continue to advance our thinking on these matters. Chapter 5 explores these works and pinpoints a way forward, providing a way to develop a guided method to address topics that involve multiple groups that require moral consideration. This investigation culminates by showing that the only approach equipped with the necessary structure to address such dilemmas *in full* is weak anthropocentrism. Through providing this picture, we will see how we can take the fundamental pattern of structural ethics and expand it to maximize its utility, which will bolster the ability to see the full range of transportation's impacts in their entirety. After unpacking these ideas, we see how thinking in such a manner advances our capabilities to work toward worthwhile goals such as socially just urban sustainability through starting with transportation systems. As a reminder, as mentioned during the critique of Jonas's imperative, attending to such matters will help the people who need it the most.

NOTES

1. I explore this idea in a different context elsewhere. For more information, see Shane Epting, "Urban infrastructure and the problem of moral praise," *Techné: Research in Philosophy and Technology* 25, no. 2 (2021): forthcoming.

2. To see a more robust account of the many kinds of possibilities that fit under this description, see Edward Glaeser, *Triumph of the city: How our greatest invention makes us richer, smarter, greener, healthier and happier* (New York City: Penguin, 2011).

3. It is worth mentioning that I understand that prioritizing the attention that ethically-based affairs receive is a position that could indeed challenge the view that I present. That is to say, although I am focusing on transportation and the need to increase the attention that it receives from ethical theorists, one could challenge this view. They could hold that any number of applied issues deserves prioritization. This challenge is formidable. However, engaging in such a discussion does not take away from the reality that mobility issues have prioritization concerns that are central and inherent to their situatedness. As for the meta-level concerns, such issues fall far outside of the current project and deserve consideration at a latter date.

4. Martin Luther King Jr, "Letter from Birmingham jail," in *Liberating faith: Religious voices for justice, peace, & ecological wisdom*, ed. Roger Gottieb (Lanham, MD: Rowman & Littlefield Publishers, 2003), 177–87.

5. Jonas, *Imperative*, 10.

6. For a few examples, see Bus Riders Unite, http://www.opalpdx.org/bus-riders-unite/ and Rainier Beach Transit Justice Project, http://www.rbcoalition.org/rainier-beach-transit-justice-youth-corp-completes-metro-mural-their-latest-project-to-better- our-community-transportation-wise/.

7. It is worth mentioning here that philosophical commentary of the actual arrangement of parts would be best accompanied or followed by professionals who have the training to identity the kinds of concerns that are above a philosopher's pay grade. These notions include but are not limited to functionality and safety.

8. It is worth mentioning that other forms of transportation design can also take away from walkability. For more information on this point and the idea that automobile-centric planning can affect a city's walkability, see Ann Forsyth *et al.*, "Cities afoot—Pedestrians, walkability and urban design," *Journal of Urban Design* 13, no. 1 (2008): 1. For a specific reference that zeroes in on the interplay between automobile-centric design, walkability, and health, see Lawrence Frank *et al.*, "Many pathways from land use to health: associations between neighborhood walkability and active transportation, body mass index, and air quality," *Journal of the American Planning Association* 72, no. 1 (2006): 75.

9. Juliana Maantay, "Asthma and air pollution in the Bronx: Methodological and data considerations in using GIS for environmental justice and health research," *Health & Place* 13, no. 1 (2007): 47.

10. Giulio Mattioli, "Forced car ownership" in the UK and Germany: Socio-spatial patterns and potential economic stress impacts," *Social Inclusion* 5, no. 4 (2017): 148. The literature on this topic shows that forced-car ownership is controversial. However, the aforementioned article provides a thorough representation of such debates.

11. Philip Brey, "From moral agents to moral factors: The structural ethics approach," in *The moral status of technical artefacts*, eds. Peter Kroes and Peter-Pauk Verbeek (Dordrecht, NL: Springer, 2014), 125. I also examine some of these points in a different context here: Shane Epting, "On municipalities as technologies," *Philosophy & Technology* (2021): 1–11, https://doi.org/10.1007/s13347-020-00438-z.

12. Bruno Latour, "Where are the missing masses?, The sociology of a few mundane artifacts," in *Shaping technology, building society: Studies in sociotechnological change*, eds. Wiebe Bjiker and John Law (Cambridge, MA: MIT Press 1992), 225. Brey makes this point also, in Brey, "Moral Agents," 128.

13. Latour, "Missing Masses," 226.

14. Ibid., 227; 244ff.

15. Brey thoroughly examines and charts the trajectory of these arguments following Latour. For more information, see Brey, "Moral agents," 128ff.

16. Peter-Paul Verbeek, *Moralizing technology: Understanding and designing the morality of things* (Chicago: University of Chicago Press, 2011), 11ff. There points are also examined here: Shane Epting, "On Municipalities."

17. Verbeek, *Moralizing*, 38–9.

18. Verbeek, *Moralizing*, 61.

19. Brey, "Moral agents," 134.

20. Brey, "Moral agents," 135.

21. Karel Marten, *Transport justice: Designing fair transportation systems* (Abingdon: Routledge, 2016), 37.

22. Matt Doeden, *Choppers* (Minneapolis, MN: Lerner Publications Co., 2008), 15–16ff.

23. Brey, "Moral agents," 126.

24. It is worth underscoring here that "moral parts" does not entail that they have a traditional sense of morality that requires discussions about topics such as agency and intentionality.

25. My research and thinking into this area have evolved over the years. I had previously referred to this process as "complex moral assessment" in several places. For a few examples, see Shane Epting, "On moral prioritization in environmental ethics: Weak anthropocentrism for the city," *Environmental Ethics* 39, no. 2 (2017): 131–46. Also, see Shane Epting, "Automated vehicles and transportation justice," *Philosophy & Technology* 32, no. 3 (2019): 389–403. However, I now see that thinking in a manner that focuses so strongly on evaluations is shortsighted in the sense that it wants to produce a kind of rough "checklist." The problem with this approach is that it an assessment does not favor the kind of flexibility that a process such as moral ordering provides. An assessment implies that there is a kind of strictness at play, which pushes against the "anti-framework" framework. Moral ordering, however, does not bear this connotation, and, if it does, it does so to a lesser degree because it has more of a *suggestive* character. This quality is necessary for dealing with situations that differ with respect to any number of elements that could come into view in a given city.

Chapter 5

The Pathway to Moral Ordering

Despite the endeavor to address mobility problems that are almost exclusive to contemporary cities, the thinking underpinning such an approach is not new. There is a lengthy history in philosophy that deals with issues that involve more than one, if not several, categories that require moral consideration. Such dynamics suggest that there have been and will continue to be occasions wherein one party deserves to be morally placed higher in decision-making, leading to actions. These steps inevitably show a preference for one grouping over another, even though such placements are not permanent or strongly universal. Still, it seems reasonable to strive toward widely applicable guidelines to build onto results that work in our favor when undertaking a similar endeavor. This notion gestures toward the universal, but to reiterate a point made in chapter 1, it is only an element that happens to align with universalizability. To be clear, it is not an advocation of a specific worldview, just a proposed method for dealing with decisions wherein numerous elements require "moral balancing."

The idea worth bearing in mind when considering the desirability of such a view is that it is geared toward issues, specifically in urban mobility, that demand flexibility. Such a claim indicates that urban problems' unique nature could benefit from philosophical insights that the canon (along with its neglected approaches) can provide. This kind of attitude is conducive to dealing with problems wherein fluid conditions hold steady, which is a characteristic commonly associated with cities. If developed according to the above specifications, we gain a methodological way of thinking that provides guidance for addressing ethical affairs that strive for elements such as universalizability, repeatability, and predictability. Yet, it also allows for necessary adjustments that operating on a case-by-case basis necessitates. Although these kinds of situations are not always troublesome, instances that need attention are bound to occur in transportation. Such cases require that we study their moral dimensions. For these scenarios, deciding on one option over another becomes a moral decision. I refer to such situations as instances of the problem of moral prioritization.[1]

Framing these scenarios in this manner is advantageous because it highlights the moral crux of decisions that involve multiple stakeholders. The problem of moral prioritization provides a way to set up the issue to examine which stakeholders require consideration, allowing us to discover the moral weight that each grouping carries. It gives structured guidance for the tough calls that must be made when transportation specialists cannot make decisions that appease everyone. This device would be incredibly useful if one category must endure hardship while another group benefits, and there is the need to explain *honestly* why it has to work out that way. If such situations occur, then additional efforts should develop amicable avenues that can improve a bad outcome, rather than merely accepting it.

The significant challenge with making such calls is that it opens up room for abuse, making a decision wherein a category advances the degree of consideration they receive while a more deserving group suffers. That is, often, decisions that involve transportation infrastructure come with price tags in the millions or, in some cases, billions. Attempts to channel such funds through influencing decisions must not be discounted. Although we should not disregard this unfortunate reality, it is also not an absolute condition. Yet, acknowledging this possibility proactively must remain at the forefront of the decision-making process. One can argue that keeping it in mind and part of the discussion about such decisions could reduce the likelihood of its occurrence.

While employing this approach in terms of transportation requires that we deal with numerous categories that need moral ordering, utilizing it is also progress for the philosophical enterprise of multilevel-stakeholder engagement. Perhaps one of the most well-known and early examples of this dynamic is derived from Jean-Paul Sartre's existentialism, wherein he examines the case of a man who had to choose between attending to his sick mother or joining the military efforts of World War II.[2] Baird Callicott made extensive use of this case, showing how it could serve as a precedent model for making decisions between groups of stakeholders who vastly differed.[3]

Callicott focused on the idea that, although he recognized and accepted that he had a duty to care for her, the man also was aware that he had a more substantial obligation to defend his country in a time of great conflict.[4] After weighing the stakes, the man left his mother to support the war effort to fight in battle. He made this difficult choice because he recognized that his obligation to his country, including protecting his mother from enemy forces, outweighed the duty to care for her during her illness, even though she needed immediate care.[5]

Although one can argue that not all cases that resemble the above will require the same or similar actions, from this example, we can think about it in a way that illustrates how the real-world demands that one must make difficult choices wherein the immediate answer is not apparent. Challenging decisions can call for actions that require a well-thought-out response that appear to go against a more pertinent interest. However, once one sees that there are good reasons for changing the order of consideration, the motivation behind such decisions is clear. Although the case above reveals that conflicting moral obligations—having to choose between one stakeholder, the mother, and another stakeholder, the country—remain within the sphere of human-to-human concerns. Yet, the pattern behind such decisions has been useful in other areas that involve nonhuman stakeholders.

For instance, Callicott employs Sartre's above dilemma when exploring ways to approach competing interests between humans and nonhumans when there is a situation that lacks an obvious answer. He shows that there are cases wherein humans and nonhumans are at odds, and one stakeholder must prevail in a way that the competitor is at a disadvantage. Matters become complicated because there are two primary kinds of values at play, intrinsic and instrumental. These notions make up the grounding that environmental philosophers such as Callicott and Holmes Rolson stand on when arguing that humans must shift viewing ecosystems as a means to an end—to an end in themselves, considered as ecological wholes; a view called ecocentrism.[6] The essential benefit that it gives us is bringing these values into perspective, being able to advance arguments for nonhumans that do not succumb to mere economic considerations in every instance.

Despite advancing the conversation from human-to-human affairs to human-to-nonhuman matters and bringing the above values into the forefront of our thinking, Callicott's ecocentrism has limits and conceptual issues that will prove to be difficult for addressing problems that ultimately concern humankind, such as urban mobility. That is, transportation is a human-centered matter. In turn, it requires an approach that keeps the moral focus on humanity. However, this notion does not dismiss the idea that the nonhuman world deserves moral consideration, aside from any competing interest from the human world.

Through analyzing this position, we see that the benefit of Callicott's approach for dealing with transportation is that he advances the conversation in terms of the diversity of stakeholders within the problem of moral prioritization. However, we must confront and overcome the conceptual limitations integral to his version of ecocentrism before moving forward. To account for these conditions, the following section examines how Callicott and other environmental philosophers situate ecocentrism. Next, I show how the many problems with it require us to abandon any hope of applying it to transportation problems centered on humankind. With these aspects included in how we see the history of multi-tiered stakeholder engagement, we gain a view showing how the following chapters that explore moral ordering not only provide a way to think about complex problems in transportation planning and engineering, but they also exhibit how engaging in such thinking benefits how we understand issues that force us to consider different types of stakeholders.

EXPLORING ECOCENTRISM AND ITS CHALLENGES FOR TRANSPORTATION

It is worth mentioning that in general, I resist addressing the reasons why we should not busy ourselves by arguing why "this-framework" or "that-framework" is ill-equipped to deal with transportation issues. There are so many frameworks in the history of philosophy and other fields that such an undertaking would require a separate book. Such a task makes about as much sense as test driving all the vehicles you do not want to purchase, only to show someone that the car you wish to drive runs well, especially for specific tasks. Such an attitude feeds the assumption that philosophy should *only* compete in the marketplace of ideas instead of existing for its own sake. The former relies heavily on hyper-competition while the latter bolsters the purity of the philosophical enterprise. This (somewhat romantic) point aside, when there is a particular one of a very similar make and model, it is reasonable to examine it, just to know why you are secure in your purchase. Ecocentrism is one such case.

It holds steady as the most developed approach within the literature that makes use of intrinsic value for dealing with conflicts between humans and nonhumans, which we must consider if we want to examine the full range of values at play. The general tenets of this position maintain that ecosystems' intrinsic value should motivate our decisions behind environmental consideration.[7] In turn, ecocentrism should be well primed for addressing multifaceted urban concerns, and transportation issues would naturally fit the bill. The way that Callicott unpacks the mechanics behind ecocentrism illustrates how it advances (what I call) moral prioritization. He exhibits the inherent conflicts that arise when they go beyond the dynamics that he fleshed out from the case that he took from Sartre. His goal is to show that to overcome the thinking that has led to the kinds of outcomes that entail massive ecological degradation, humans must shift their mindset completely, beyond mere immediate and narrow interests.[8]

The thinking behind this view largely rests on Aldo Leopold's seminal work, *A Sand County Almanac: Sketches Here and There*, encapsulated in the following excerpt: "The land ethic simply enlarges the boundaries of the community to include soils, waters, plants and animals, or collectively: the land. . . . In short, a land ethic changes the role of *Homo sapiens* from conqueror of the land-community to plain member and citizen of it."[9] Inspired by Leopold's thought, several environmental ethicists have developed elaborate systems to address humankind's place in the world.[10] For example, Callicott explains that ecocentric thinking emerged when environmental philosophers called for a complete overhaul of Western thought, moving the locus of intrinsic value from people to entire ecosystems.[11] The goal was to build a new metaphysical and ethical paradigm for thinking about the world.[12]

The call to re-center how humankind fits within the continuum of life, as a member of a larger whole, is the essential criterion that defines the ecocentric position, a notion justified by non-anthropocentric intrinsic value. Establishing the difference between instrumental and intrinsic values provides a view showing that we have reasons to examine a broader range of considerations for nonhuman life than taking inventory of what it can do for us. Yet, Eugene Hargrove fleshes out intrinsic value, giving us the motivation to see fully why this issue deserves attention for philosophical and practical reasons. He identifies two kinds of such values in the literature, objective non-anthropocentric and subjective non-anthropocentric.[13] The former holds that nonhumans have intrinsic value that exists without humans valuing them, while the latter relies on valuing by humans.

Holmes Rolston endorses the objective non-anthropocentric view of intrinsic value.[14,15] Per his formulation of this approach, he argues that all nonhuman life has intrinsic value, just as they have other characteristics.[16] This kind of intrinsic value is comprehensively understood through thinking about each living thing as it exists in an ecosystem; all life depends on interconnected relationships, but each organism strives to maintain itself.[17] Within this web, each organism requires the instrumental value of other organisms for metabolism.[18] Because they defend their lives in such a manner, they have intrinsic value, respectively.[19] Due to this assemblage, these organisms, existing as an ecosystem, have systemic value.[20,21] Considering that they have this value, they carry moral weight, and we have a duty to respect it.[22]

Slightly differing from the objective view, non-anthropocentric subjective intrinsic value maintains that nonhuman life does not have intrinsic value unless humankind values it.[23] Callicott favors this position, and he shows how it applies to environmental conflicts of interests between human and nonhuman problems.[24] For instance, he lays out his second-order principles [SOP-1 and SOP-2] as ways to deal with conflicts of interest between human and nonhuman considerations. Callicott maintains that membership in an immediate community gives a person SOP-1 obligations that outweigh distant communities' concerns.[25] For example, a person has an SOP-1 obligation to their family that overrides a responsibility to their neighbors. SOP-2 obligations maintain that a person has stronger interests toward duties that outweigh weaker interests.[26] Consider, for instance, that I have an obligation to spend time with my family (SOP-1), but if I do not earn money at my job, my family will be homeless. In turn, my SOP-2 obligation has more importance than my SOP-1 obligation.[27]

The SOP-1 and SOP-2 arrangement provides a blueprint for how to justify decisions that prioritize nonhumans over humans. The strength of this position is that it allows humans to put their interests ahead of nonhumans. Still, there is some flexibility when stronger nonhuman interests arise that require re-prioritization. Due to its structure and flexible nature, Callicott's ecocentrism provides a means to make environmentally minded decisions without moral conflicts, supposedly. Ecocentrists could argue that these qualities make it well suited as a problem-solving measure. For instance, Callicott exhibits how these principles work:

> The spotted owl is threatened with preventable anthropocentric extinction—threatened with biocide, in a word—and the old-growth forest biotic communities of the Pacific Northwest are threatened with destruction. . . . If we faced the choice of cutting down millions of four-hundred-year-old trees or cutting down thousands of forty-year-old loggers, our duties to the loggers would not take precedence by SOP-1, nor would SOP-1 be countermanded by SOP-2. But that is not the choice we face. The choice is between cutting down four-hundred-year-old trees, rendering the spotted owl extinct, and destroying the old growth forest biotic community, on one hand, and displacing the forest workers in an economy that is already displacing them through automation and raw-log exports to Japan and other foreign markets. And the old-growth logging lifestyle is doomed, in any case to self-destruct, for it will come to an end with the " final solution" to the old-growth forest question, if the jack-booted timber barons continue to have their way. With SOP-2 supplementing SOP-1, the indication of the land ethic is crystal clear.[28]

From the example above, we can extrapolate a guide for how to put ecocentric principles into practice. Although Callicott is addressing wilderness preservation, applying his second-order principles to cities can inform us about how urban elements should or should not affect ecosystems. We could easily put SOP-1 and SOP-2 into mereologically inspired terms, analyzing how we ought to prioritize consideration for human and nonhuman parts. This notion exhibits that designing a way to think about multi-stakeholder engagement could easily rely on ecocentrism to provide the means to deal with issues that remain associated with urban mobility and the nonhuman world.

For instance, consider highways versus mass transit. One could argue that public transport parts such as light rail and buses have a lesser impact on the nonhuman environment, whereas highways and automobiles harm several species and exacerbate climate change. Each person should have the "right" to safely drive as she, he, or they see fit, but moral obligations to ecosystems are of greater importance. Consequently, city leaders should advocate on behalf of mass transit, an ecocentric defense of urban mobility.

This example exhibits how to employ Callicott's methods in a manner that is consistent with Leopold's land ethic. It shows how urban dwellers can reduce their environmental footprints by selecting infrastructures that minimize resource consumption and carbon emissions. However, despite its numerous strengths and reliability, as a sound position with a noble underpinning, there are a few reservations that we should hold that might prevent us from fully advocating for this route. For instance, ecocentrism has some conceptual criticisms that could affect its integrity and applicability, especially when dealing with affairs such as urban mobility that seem to be inherently anthropocentric. In the next section, I examine these issues to discover why such shortcomings could require us to search for an alternative approach to address such concerns.

PROBLEMS FOR ECOCENTRISM FOR TRANSPORTATION

Although ecocentrism has an honorable underlying motivation, a well-thought-out foundation, and a means to employ its principles, both versions (as mentioned above) have challenges. For the objective non-anthropocentric–based approaches, severe criticisms in the history of environmental ethics indicate why they might not be the best-equipped approach for dealing with cities. These reasons, of course, remain applicable for much of the scope of transportation affairs. For instance, Hargrove points out that they *require* anthropocentrism: "After discovering that something has a good of its own, the human or humans must decide to intrinsically value it."[29]

Through making this claim, Hargrove shows that even if objective non-anthropocentric intrinsic values do exist, acting on their behalf cannot occur unless one does so as a human who is capable of valuing. The problem here for ecocentrism is that it was supposed to replace anthropocentrism, reorienting our view of the nonhuman world wherein we are simply members of the biotic community. Yet, anthropocentrism is a necessary component of objective non-anthropocentric thinking and acting, suggesting that it is a troubling concept.

While this criticism takes aim at objective non-anthropocentrism, it also indicates a problem for subjective non-anthropocentric intrinsic value. For instance, Hargrove argues that we cannot escape having a human's perspective, alluding to Thomas Nagel's "What's it like to be a Bat?" Hargrove illustrates that the problem with non-anthropocentrism rests with our inability to access the required knowledge to fully experience being a bat (or any other nonhuman living thing).[30,31] If such an experience is impossible, then describing aspects that stem from such experience rests on guesswork. Replacing "bat" with "ecosystem" epistemologically complicates matters because there is no way to know what it is like to be a mindless system of different species. Echoing Hargrove, Frederick Ferré formulates this condition as perspectival anthropocentrism, while other philosophers refer to this view as *conceptual anthropocentrism*.[32,33] Even if one were to appeal to teleological interests, every species would need this consideration, which then requires us to weigh such interests against each other. In turn, this argument would be incredibly thin, and it does not escape the epistemological challenge mentioned above. Yet, apart from these criticisms, there are practical concerns for ecocentrism's inability to deal with urban issues such as transportation.

For example, Rolston's ecocentrism does not make significant room for the abiotic community's intrinsic value (in the Hargrovian sense).[34] In turn, his approach cannot significantly address issues that pertain to environments such as caves. If he cannot address nonhuman nonliving entities with gusto, then weighing in on metropolitan areas' intrinsic value would not fare any better. For instance, when deciding on whether to make room for additional highway lanes that would bring the historic preservation of architecture into the conversation, objective non-anthropocentric intrinsic value has nary an application here. If we want to appeal to the kind of intrinsic value that Hargrove supports for thinking about topics such as architecture, then we are at a loss.

When it comes to applicability problems, Callicott's version of ecocentrism also raises concerns for its practicality, a consideration that weighs heavily for developing solutions to real-life transportation problems. For instance, regarding SOP-1 and SOP-2, as instruments for addressing ecological affairs in cities, it is challenging to think that there could ever be a case wherein humankind would or could act against its interests unless dire consequences would show otherwise, especially considering Hargrove's criticisms about the mere possibility. For example, when using SOPs to make a complex decision that involves balancing human and nonhuman interests, it does not make sense to say that nature could have an interest in anything beyond essential biological well-being and flourishing. These very notions cannot escape their anthropocentric-situated conditions.

What does an ecosystem have to say about transportation infrastructure? The only answer that can come with any degree of certainty would be identifying what counts as being in humankind's direct or indirect (long-term environmental) interests. If this notion holds, then it means that the thinking behind SOP-1 and SOP-2 remains anthropocentric, yielding human-favored outcomes. In turn, ecocentrism fails to provide a way to work in practice that is true to its principles. This notion suggests that it cannot reorient our thinking about the nonhuman world in a congruent manner with cities on epistemological or practical grounds.

Aside from these criticisms, ecocentrism faces other challenges when it comes to avoiding issues that rest on its inherent structure—which also apply to all of the foundational approaches in Western academic environmental ethics. The problem is that all such approaches fail to prioritize humans above nonhumans explicitly. This orientation suggests that human-to-human social-justice concerns are not strong motivating considerations when stacked against nonhuman ones. In all fairness, neither Callicott nor Rolston gave any indication that they were developing an approach to social justice, transportation justice, or any kind of other value-laden enterprises that falls beyond an ecological purview. This rebuttal would be excellent, but ecocentrism, along with all Western academic environmental ethics, is still a moral theory, meaning that it is not exempt from moral scrutiny. This point is not to suggest that ecocentrists or environmental philosophers go against any kind of social-justice causes—in any shape, form, or fashion to any degree whatsoever. The fact that these philosophers have placed such high importance on the well-being of the nonhuman world shows that they have noble aspirations that deserve praise.

However, the field of environmental philosophy and thought has had its share of criticisms. They indicate that some fine-tuning might benefit a panoramic view of global values in one sense and that examining other instances with a handheld lens might reveal inherent limitations that we need to go beyond to see how pro-environmental thought can be compatible with the criticisms below. Mentioning these notions are not done to vilify ecocentrism, but they show that environmental philosophy, along with much of the canon of Western academic thought, needs revisiting.

For instance, on the global scale, Ramachandra Guha is critical of environmental ethics (specifically mentioning deep ecology as it was developed in the United States), arguing that discussions in the traditional literature support imperialism.[35] Bill Lawson points out that some early works in environmental ethics that dealt with cities fail to consider the poor people who live in them.[36] Charles W. Mills holds that environmental works neglect to account for the political realities of race and place.[37] Having served as the editor in chief of *Environmental Ethics* since 1979, Hargrove argues that environmental ethics has mostly neglected environmental racism, pointing out that most philosophers mainly research subjects such as protecting ecosystems and nonhumans .[38] Considering these criticisms as a collection, they paint a picture showing that, by concentrating on some problems, environmental thinkers and philosophers unintentionally neglect such issues. The points above are not meant to suggest that ecocentrism is not the only approach within environmental philosophy or thought that must shoulder the above contentions.

Moving forward, the path should be clear. The aim should be to act so that our actions reflect our moral priorities, and weak anthropocentrism might provide a consistent, moral way to address complex urban affairs such as transportation infrastructure and policy in such a manner. This point does not entail that weak anthropocentrism is structurally anti-racist, anti-classist, anti-sexist, anti-ablist, or anti-ageist. From what has been presented so far, there is no reason not to include it in the criticism of approaches mentioned above. As a position within Western mainstream environmental philosophy, it remains subject to all of those criticisms. However, its structure provides the opportunity to put significant distance between itself and other environmental theories in the discipline. With some elbow grease, showing how this claim holds will show why we should champion this approach for such reasons.

What is more, we will investigate how its inherent nature is compatible with structural ethics, a quality that will allow us to maintain the direction that this project is going. It began by showing how we can talk about transportation systems as wholes that have numerous smaller parts. It has now advanced to examining the morality of arranging parts in a particular manner that brings the context of such affairs into view. The goal here is to help create the conditions for dealing with the inherent sensitivities that come with real-world issues in urban mobility. The attention is turning to how we can include nonhuman parts that have intrinsic value into the conversation that does not dismiss the moral dimensions of such arrangements. The idea that we want to keep firmly in view is that, as mentioned previously, to develop a manner of contemplation that focuses on moral ordering for dealing with these kinds of affairs, one that can deliver transportation systems that support worthwhile goals such as socially just urban sustainability. In chapter 6, I examine these ideas, searching for a way to balance the many relevant considerations morally.

NOTES

1. Shane Epting, "On moral prioritization in environmental ethics: Weak anthropocentrism for the city," *Environmental Ethics* 39, no. 2 (2017): 131.

2. Jean-Paul Sartre, *Existentialism,* trans. B. Frechtman (New York: Philosophical Library, 1947), 28–32.

3. J. Baird Callicott, *Beyond the land ethic: More essays in environmental philosophy* (Albany, NY: SUNY Press, 1999), 73–4.

4. Callicott, 73–4.

5. Ibid.

6. Hargrove notes these views as non-anthropocentic, which is ecocentrism broadly. See Eugene Hargrove, "Weak anthropocentric intrinsic value," *The Monist* 75 no. 2 (1992): 184.

7. For an example, see J. Baird Callicott, *In defense of the land ethic: Essays in environmental philosophy* (Albany, NY: SUNY Press, 1989), 3–4.

8. Callicott, *In defense*, 3–4.

9. Aldo Leopold, *A Sand County Almanac: Sketches here and there* (New York: Oxford University Press, 1949, 1968), 204.

10. Hargrove, "Weak," 183.

11. Callicott, *In Defense*, 3–4.

12. Ibid.

13. Hargrove, "Weak,"184.

14. Ibid.

15. In is worth mentioning that Katie McShane holds that Rolston's view of intrinsic value endorses the following: "Views according to which claims about the intrinsic value of X are claims about which properties of X make it valuable. . . . Views according to which claims about the intrinsic value of X are claims about the metaphysical status of X's value properties." See, Katie McShane, "Why environmental ethics shouldn't give up on intrinsic value," *Environmental Ethics* 29, no. 1 (2007): 47.

16. Ibid.

17. Holmes Rolston, *Environmental ethics: Duties to and values in the natural world* (Philadelphia: Temple University Press, 1988), 100.

18. Holmes Rolston, *Conserving natural value* (New York City: Columbia University Press, 1994), 177.

19. Rolston, *Conserving*, 177.

20. Ibid.

21. Rolston, *Environmental ethics*, 186–7; also see 216.

22. Ibid., 230–1.

23. Hargrove, "Weak," 194.

24. J. Baird Callicott, *Beyond the land ethic: More essays in environmental philosophy* (Albany, NY: SUNY Press, 1999), 73–5.

25. Callicott, *Beyond*, 73–5.

26. Ibid.

27. Kristin Shrader-Frechette develops a similar approach to address issues between human and nonhuman interests, coining the term "hierarchical holism" to provide an elementary account of moral prioritization. For more information, see Kristin Shrader-Frechette, "Individualism, holism, and environmental ethics," *Ethics and the Environment* 1 (1996): 55–69.

28. Callicott, *Beyond*, 75.

29. Hargrove, "Weak," 191.

30. See Thomas Nagel, "What is it like to be a bat?," *The Philosophical Review* 83, no. 4 (1974): 435–50.

31. Hargrove, “Weak,” 201.

32. Frederick Ferré, “Personalistic organicism: Paradox or paradigm?” *Royal Institute of Philosophy Supplements* 36 (1994): 72.

33. See Allan Thompson, “Anthropocentrism: Humanity as peril and promise,” in *The Oxford handbook of environmental ethics*, eds. Stephen Gardiner and Allen Thompson (New York: Oxford University Press), 82.

34. Hargrove, “Weak,” 193–4. Also, see Rolston, *Environmental ethics*, 232–45.

35. Ramachandra Guha, “Radical American environmentalism and wilderness perservation: A third world critique,” *Environmental Ethics* 11, no. 1 (1989): 76.

36. Bill Lawson, “Living for the city: Urban United States and environmental justice,” in *Faces of environmental racism: Confronting issues of global justice*, eds. Laura Westra and Bill Lawson (Lanham, MD: Rowman & Littlefield Publishers, 2001), 41.

37. Charles Mills, “Black trash,” in *Faces of environmental racism: Confronting issues of global justice*, eds. Laura Westra and Bill Lawson (Lanham, MD: Rowman & Littlefield Publishers, 2001), 74.

38. Eugene Hargrove, “Forward,” in *Faces of environmental racism: Confronting issues of global justice*, eds. Laura Westra and Bill Lawson (Lanham, MD: Rowman & Littlefield Publishers, 2001), ix.

Chapter 6

Moral Prioritization in Urban Mobility

Recalling a point from previous chapters, transportation planning decisions affect vulnerable and marginalized people, the public, nonhumans, future generations, and historically and culturally significant artifacts. All of these groups deserve consideration—but not equally. This point holds that choices pertaining to the order of whom or what should receive consideration are subject to moral examination and debate. How we arrange, add, modify, or take away parts in a transportation system will change outcomes, and we need to ensure that such actions are appropriate.

For instance, weighing too heavily in favor of nonhumans while disenfranchised groups suffer could bring charges of environmental discrimination. Such a position could be tantamount to saying that green lives matter more than black, brown, disabled, or senior lives. We must avoid such egregious missteps. Employing the conceptual device of "moral ordering," a measure that can help us address the moral obligations to the categories above, can lessen the chances of acting in a discriminatory fashion.

The purpose of this chapter is to show why we need such a measure to underpin arguments that deal with multifaceted mobility issues. To make this case, I examine weak anthropocentrism because it uses the concept of intrinsic value. This quality is required to flesh out the moral dimensions of transportation affairs in the contexts of their global and intergenerational reaches. Even though I champion this position, it is not exempt from criticisms. To understand how it can avoid the discriminatory charges above, I reveal how specifying its structure's conditions can avoid unintended prejudice. That is to say, as a framework, weak anthropocentrism can be customizable, which allows it to adapt to specific circumstances. Bearing in mind that it lacks a rigid operational structure, it aligns with the description of an "anti-framework" as examined in earlier chapters. The qualifier "weak" might give one the impression that it lacks integral strength, but that point fails to consider that one of weak anthropocentrism's assets is that it makes room for "extra-human" elements, to use Jonas's term.[1] Through providing a way to take on moral considerations for the intrinsic value of nonhuman entities, weak anthropocentrism can inherently balance what would appear to be competing interests.

In turn, this chapter goes in that direction. It builds onto the ideas explored on mereology-inspired concepts and structural ethics, showing how we can take the theoretical notions that began with parts and parthood relations, using that structure to guide parts' arrangement to yield moral outcomes. While these points are straightforward, dealing with at least five groups of stakeholders challenges the process, practically and conceptually. Weak anthropocentrism, I show, attends to both concerns. It gets things moving.

The point worth underscoring here is that there are some conflicts of interest among different categories of entities, such as human and nonhuman, which are not *actually* competing. Recalling Hargrove's wisdom, all such interests are rooted and epistemologically grounded in the undeniable anthropocentrism that makes valuing and knowledge of it possible. Due to this reality, the only interests that are "competing" are human interests that we must prioritize, even though our interests can involve nonhumans. In turn, we are dealing with outcomes that stem from how the parts of a transportation system and their arrangements affect humankind, but this notion does not mean that nonhuman nature is entirely free for the taking or mindless destruction.

This condition not only makes it compatible with structural ethics, but it complements it rather nicely, serving it in an extensive fashion that gives us a way to address how outcomes affect the categories mentioned above. Weak anthropocentrism is congruent when connected to structural ethics, serving as the next step in guiding our transportation decisions. Thought about in one of these two ways, weak anthropocentrism shows us the first reason why we have to make a rough distinction early on in the process of moral ordering. That is, humans ought to come first when moral ordering, a notion that aligns with our epistemologically situated condition that we inherit from Hargrove. The problem here is this: Which humans come first when it comes to securing desired outcomes, and how do we establish meaningful criteria to make such a determination? After addressing these points, there is still a case that nonhuman considerations should come into view because we have an interest in them for their intrinsic and instrumental values. Yet, there is also the need to address future generations and urban artifacts' positioning, which will have to wait until chapter 7 to receive sufficient attention.

INTRINSIC VALUE AND URBAN MOBILITY

Perhaps the most relevant work to mention transportation issues in an urban context that involves an entanglement of human and nonhuman topics is Robert Kirkman's *The Ethics of Metropolitan Growth: The Future of our Built Environment*.[2] In this work, he examines a wide range of issues that show how urban sprawl is a subject of study that impacts numerous areas of life, including transportation, of course, but also education, wilderness, and community. His book deals with the "big-ticket" items that one must examine when ethically thinking about urban sprawl, issues that will affect millions of people. While his focus is not on transportation exactly, Kirkman wrestles with how metropolitan areas antagonize the nonhuman world, but he avoids intrinsic value, paying attention to policy issues instead. Although he deserves praise for his practical approach, disregarding intrinsic value narrows the scope of inquiry. In turn, we still need to provide a full account of cities' moral dimensions and their relationships with transportation.

While there are several ways to conceptualize it, the model for which I advocate fits Hargrove's view of weak anthropocentric intrinsic value. He argues that we can understand the subject by thinking about how we understand art.[3] For example, while great paintings can emotionally move us, such works rely on standards of taste that change with the times. Yet, they do not lose their intrinsic value when popular appeal dictates, but instead, they depend on experts who can make relevant social assessments.[4] When it comes to applying this line of thought to the nonhuman world, these values motivate nature "experts" and trained enthusiasts.[5]

While one might get the impression that Hargrove is confusing intrinsic and instrumental values because they both involve aesthetics, this condition is merely coextensive. Instrumental value entails something of beauty, giving the person who is valuing it pleasure. However, intrinsic value consists of an appreciation of the thing in question that requires an act in judgment that goes beyond mere aesthetic pleasure. For such assessments, worries should rest on standards and ideals used for evaluation. The danger is that people will try to demean an object, claiming that it only has instrumental value as a way to justify exploitation.[6]

This view of intrinsic value does not only pertain to the arts and environmental affairs, but library historians have made the case that we should care for old documents that have been digitally preserved because they have this kind of intrinsic value.[7] That is to say, making a digital copy of an old document preserves its instrumental value, but we can still value the original for its own sake. This pattern of thought reveals that intrinsic value is an area of concern that deserves significant consideration. I argue that it is required to deal with urban-mobility affairs that involve human and nonhuman elements.

In the next section, I examine how the weak anthropocentric position is inherently geared toward helping us think about such issues. It does so in a manner that is consistent with structural ethics, serving in an auxiliary capacity.

WEAK ANTHROPOCENTRISM FOR TRANSPORTATION

Despite having thoroughly developed his version of weak anthropocentrism to address human and nonhuman affairs, Hargrove was not the first philosopher to employ this approach. For instance, although he does not appeal to intrinsic value, Bryan Norton developed a weak anthropocentric position to argue against strong anthropocentrism, the view that humans determined all value based on the degree of satisfaction of our *felt* preferences.[8] In contrast, weak anthropocentrism holds that humans' *considered* preference determines all values.[9] This idea suggests that our well-reasoned thinking about the environment could lead to environmental protection.[10] For example, we could argue for environmental protection due to an interest in having thriving ecosystems for future generations, a move that speaks to efforts in sustainability. In turn, Norton shows how humankind can develop approaches to environmental consideration that do not bring charges of misanthropy.

This view benefits environmental ethics because it provides a way for humans to include considerations for nonhumans, but it receives criticism because it neglects intrinsic value and excludes other dimensions such as aesthetics.[11] Norton has since abandoned this position, opting for a pragmatic approach instead.[12] Hargrove maintains a different version of weak anthropocentrism, focusing on intrinsic values, as described earlier. For him, dismantling non-anthropocentrism (e.g.,ecocentrism) involves showing how it is a version of weak anthropocentrism.

Bearing in mind his criticisms of non-anthropocentric intrinsic value mentioned in chapter 5, we understand how he grounds his approach. For instance, while Hargrove holds that we cannot escape having a human's perspective, appealing to nonhuman (living and nonliving) intrinsic beauty to justify moral consideration of the nonhuman world provides a means to defend it.[13] Through including this stipulation in his version of weak anthropocentrism, he goes beyond Norton's preferences/consideration framework. This move also justifies moral consideration based on intrinsic value that benefits how we understand cities' beauty and their transportation systems.

For instance, Hargrove's work reminds us that anthropocentrism simply means "human-centered." One way to interpret the term "weak" is to suggest that anthropocentrism is flexible, showing that we can engage in practices that fall outside our immediate interests in favor of interests that bring the nonhuman world into our perspective. Weak anthropocentrism still makes humans the prime focus for consideration, but making room for nonhumans' intrinsic value makes it weak. Through making this modest concession, weak anthropocentrism can avoid missteps while dealing with the problem of moral prioritization by not placing humans beneath nonhumans in most instances. This basic structure shows that humans' interests remain respected as a highly regarded priority, and nonhumans remain in view for reasons that go beyond gross instrumentalism because they are seen as an end.

Further, recalling a point from Callicott's work on Sartre, one could add that humans also have moral obligations toward caring for their own species before caring for nonhumans. Beyond the view that this reason aligns with the suspected trajectory of moral ordering given above and is coming below, which is merely a coextensive quality, this motivation is coupled with the epistemological grounding that Hargrove establishes. This notion means that while we can only act from a human-centered position with any substantial degree of certainty, declarations that require testable and verifiable knowledge of collective nonhuman minds and some of their interests—an incredibly challenging task when dealing with an ecosystem—that could bolster many claims about morality and desired outcomes remain tethered to our inability to escape this reality. In turn, due to this lack of certainty and the inability to speak for ecosystems with absolute authority, our epistemological-situatedness can also support moral claims and their positioning in a process of moral ordering.

This idea is straightforward: we have access to humankind's interests, and we have access to the interests of groups and individuals. The challenging aspect is attaining it. Still, the possibility exists with a higher degree of certainty because we can access and verify the required information to make decisions that favor our interests. This point means that, when making decisions about mobility and humans' interests, it is possible and available. Bearing these aspects in mind, decisions that pertain to how we arrange the parts of a transportation system for ourselves come with a higher degree of certainty. Yet, even though we have access to this knowledge, such a condition does not take away from the fact that we can and do act (or have acted) against the interests and well-being of humankind, groups, and specific individuals. While tragic, this point highlights the idea that we do have the ability to act for such reasons, but the history of transportation exhibits otherwise.

For instance, in the United States, engineers have focused on the advantages of mass transit over personal vehicles for almost 100 years, basing their views on apolitical dimensions such as efficiency. Consider that in 1931, Charles Gordon, managing director of The American Electric Railway Association, noted that people's interest in personal vehicles created horrific congestion in cities such as New York.[14] This point shows that transportation authorities and residents have been dealing with mono-technical saturation for a lengthy period. Although anecdotal, this notion suggests that we cannot act or balance our interests and well-being in some cases.

Aside from this view, as discussed in the earlier chapters, the environmental impacts also support the idea that we have not secured the conditions that are in the interest of our health and well-being—even with experts in charge. Seen in this light, does the argument of not being able to act for absolute certainty for ecosystems or nonhuman life even hold? It does. Simply because we have, in many instances, failed to act for the above reasons does not entail that it is impossible. Even though it remains incredibly epistemologically challenging in the best case and enormously impossible in the Hargrovian sense, we can still act in an anthropocentric view of nonhuman interests.

Apart from this notion, although the points above show that we have no option but to prioritize humans above nonhumans, Jonas's imperative coextensively connects them. That is, he holds that we must preserve the conditions of genuine human life, meaning that nonhuman nature must exist so that humans can do the same.[15] Still, we must keep in mind that we can only do so from a position based on what we can know with the greatest certainty. In this case, it rests on what we can know about human interests—which is already a challenging subject, as mentioned. Aside from this view, one could argue that some humans want to act for nonhuman interests, even if we are projecting our interests on them. If this position is to stand, then its defenders must provide a way to balance such interests, especially when they conflict.

Although the argument above is slightly complicated, here is a summary: the argument that we have a moral obligation to humans is supported by but not fully grounded in accessible knowledge.[16] It follows that claims about ecosystems remain anthropocentric—no matter how badly we may hate the sound of it. This reason, coupled with Hargrove's practical evisceration of ecocentrism (non-anthropocentrism), shows that we are always committed to this view.

Bearing these points in mind, we see that all interests are anthropocentric, meaning that the task is to determine which interests warrant initial actions. In turn, human interests come first by default. Now, the task transitions to which humans' interests should come first when dealing with urban transportation systems. In the following section, I will argue that we can customize this element of weak anthropocentrism to speak to the situations, namely vulnerable and marginalized populations, that need be fully accounted for when arranging parts of a transportation system to produce the outcomes that we desire.

WEAK ANTHROPOCENTRISM AND MORAL ORDERING

For matters that pertain to human-nonhuman relations, weak anthropocentrism provides congruence, but examining its internal organization could reveal neglected social concerns. When it comes to thinking about transportation-related troubles, there is room to make additional moves that prioritize groups that require enhanced thinking about their place in the moral order, especially when it comes to considerations that could improve their ability to navigate urban spaces. In turn, we can make statements about the arrangement of parts that align perfectly well with structural ethics, showing how we are merely extending its utility as a process that can help us create the kind of outcomes that we desire for whole transportation systems.

Consider, for example, that during the outset of this chapter, I mentioned that failing to prioritize vulnerable or marginalized groups over nonhumans could be problematic. Structurally, weak anthropocentrism does not have any disadvantages in its design regarding social justice. With some fleshing out, it can serve as the basic blueprint that favors a just way for addressing environmental issues in transportation affairs that can escape the charges mentioned at the end of chapter 5.

For instance, Samantha Noll and Laci Hubbard-Mattix illustrate that we can advance our thinking on transportation issues by examining certain affairs through a lens of intersectionality, paying attention to gender inequalities and health disparities.[17] For the particular problems that would call for it, adding this lens to a weak anthropocentric outlook could help us see such issues more clearly. By adequately seeing specific problems in such a manner, we could improve how we morally order our mobility planning and engineering considerations. This notion suggests that transportation professionals could strategically arrange the parts of a transportation system to avoid harmful outcomes that would be subject to the criticisms that such a lens would reveal. In turn, through being able to identify exact problems, local specialists could remedy dangerous affairs that demand attention.

While this point is highly specific, it indicates that there is a larger pattern at play. It shows that the process that has been explained this far can provide a way to think about complex issues in a linear fashion, which helps structure our thinking in a well-ordered way that is consistent with the kind of means that support transferability. Through establishing even such a loose sense of this condition, we show how the quality does require adhering to a basic pattern, which is predictable and can be reproduced to some extent. Illustrating this point reveals that we are not only dealing with random aspects of transportation, but the mere possibility of absolutes existing remains possible. If we identify a pattern, it could indicate that we have found a principle that *could* be universal, but that condition is not hugely significant. Still, this notion is merely a coextensive quality of the process, not an attribute that we should dogmatically defend. For the people who are working on these affairs, this issue will matter very little when it comes to intersecting with their tasks, but these kinds of ideas do exist and are of interest to the broader philosophical conversations.

What is left to figure out for weak anthropocentrism is how to make prioritizations that properly consider these elements, which pertain to and involve competing interests as they intersect, which would matter significantly if one were examining such issues through a lens of intersectionality. If we neglect to morally prioritize the order of who or what should receive consideration, then weak anthropocentrism would not fare any better than ecocentrism, as discussed in chapter 5. Completing this task means illustrating how to embed weak anthropocentrism into moral ordering that can guide a proper and *moral* approach to moral prioritization.

To reiterate, as a theoretical device, moral ordering can help us prioritize actions that affect vulnerable and marginalized people, the public, nonhumans (individual animals and ecosystems), future generations, and urban artifacts. Each category deserves respect, but not all categories equally deserve it. I hesitate to advance a strict model because a rigid structure could result in a moral upset, wherein prioritization becomes a device for oppression. That is to say, if we think about moral prioritization as a long-term utopian goal, then committing short-term wrongs could be justified. This point again underscores Jonas's notion that we should abandon utopian visions of technology, which I add includes "philosophical devices."

Thinking in this manner suggests that we ought to frame moral ordering as a morally sound means rather than an end, but we cannot deny that achieving a moral outcome is the goal. Achieving such a reality calls for a way of thinking that can adapt to different situations, yet it is also reliable, ensuring that it promotes justice. Below, I provide a sketch of how to design for moral ordering to deal with complicated affairs in urban transportation.

Consider the following case: when debating which kind of transportation systems that cities *should* invest in, vulnerable and marginalized people, the public, nonhumans, future generations, and artifacts deserve consideration followed by action. Yet, which group should receive the highest degree of prioritization, reflected in moral ordering? I tend to side with views put forth by Emmanuel Levinas and Enrique Dussel, arguing that we ought to help people who are suffering the most, and moral ordering can support their views.[18] If one requires an additional argument for why the people who are suffering more than nonsuffering groups, Dussel holds that if we examine the sociopolitical conditions that pertain to oppression, then we discover moral reasons to ground such considerations. For instance, Dussel argues that:

> They are the ones who, by the side of the road, outside the system, show their suffering, challenging faces: "We're hungry! We have the right to eat!" That right, outside the system, is not a right that is justified by the *proyecto* or the laws of the system. Their absolute right, because they are sacred and free, is founded in their own exteriority, in the real constitution of their human dignity. When the poor advance in the world, they shake the very pillars of the system that exploits them. . . . The mere presence of the oppressed as such is the end

of the oppressor's "good conscience." The one who has the ability to discover where the other, the poor, is to be found will be able, from the poor, to diagnose the pathology of the state.[19]

From the above passage, we can infer that if the means that a politically organized society used to advance involved exploiting particular groups of people, then such conditions raise moral questions that pertain to how they should be treated in light of such histories. For the urban planner or engineer who can make decisions that can mitigate harms connected to historical mistreatment, they have a moral grounding for acting in such a manner. While they are not responsible for the origins of such harms, choosing not to act in such a fashion could perpetuate historically rooted oppression.

Due to these conditions, it seems rather axiomatic that we should aim to alleviate indirect harms that stem from a tainted past before designing a hopeful future, unless doing so can simultaneously accomplish both goals. Otherwise, such actions would make certain moral statements, ones that are not favorable, of course. When dealing with highly complex transportation systems that represent the latest advancements of technology, it seems challenging to hold that being able to deliver such an advancement could not also provide the means to right the outcomes of history's wrongs (or provide care) while "getting the trains to run on time." Failing to deliver on this point says very little about technology, but it says a lot about the people who are arranging the parts.

This point aside, there is also the challenge of knowing about such histories. To identify specific ways that a marginalized group has been unjustly treated, we can employ theoretical tools from environmental justice studies. For example, Robert M. Figueroa developed an environmental justice paradigm to pinpoint different kinds of harm. It includes physical and mental injuries, damages to cultural identity, disrespect toward traditional forms of knowledge and heritage, along with provisions for marginalized people to be included in policy decisions that affect them.[20]

While Figueroa's paradigm makes room for inclusive policy measures, I argue that the limits of participation should extend beyond such practices. To recall a point made in the previous section, we must ask: Who gets to count as being qualified to assess an urban artifact's intrinsic value? To remain consistent with Figueroa's approach, including marginalized people who have intimate access to such artifacts via cultural or historical connection is essential.[21]

The point here is that we can employ an intersectional lens such as Noll and Hubbard-Mattix's approach mentioned above, or we can appeal to principles of environmental justice.[22] The notion that is significant and deserves underscoring is that such particular theoretical devices can work better than others for unique instances in transportation. There is no good reason why we cannot design highly specialized conceptual devices that can guide our thinking and actions to deliver better outcomes for people who are suffering or whose situatedness requires specific measures. If we can address these issues and improve people's lives who have been historically oppressed, marginalized, left to fend for themselves when it comes to navigating the cityscape, or must live in ways that require care, then we should agree that such outcomes count as moral progress alongside technologically advanced transportation systems.

Bearing this point in mind, we are now in a position to return to the question, "What is a transportation system?" Recalling that the first part of the answer examines its ontological structure, then the other part reveals that it is a way to begin to deal with the myriad issues that pertain to the arrangement of parts and their moral or immoral outcomes. Put loosely, we could claim that a transportation system can deliver good and/or bad outcomes that pertain to urban mobility. Considered in such a manner, the question-and-answer, once unpacked, shows that it is a starting point for morally prioritizing the totality of harm in the city, bringing vast social and economic inequality and wide-scale environmental degradation into view. The significant dimension worth underscoring is that it does so in a manner that moves the people immediately in harm's way or disenfranchised to the forefront of our thinking. With these elements in position, we can introduce the other ones within the moral ordering process to deliver outcomes that give a us a clearer picture of what it means to pay attention to the morality of urban mobility fully.

This notion suggests that following considerations for vulnerable and marginalized people, we must then determine how a decision would harm or benefit the public, along with how they would affect the individuals. One of the most problematic situations that we find when dealing with this category within a moral order is weighing individuals' interests against the benefits of the many. That is, we need to deal with the arrangement of transportation parts that pays careful attention to this notion. In chapter 7, I will examine when this scenario becomes a problem, which is typically when these views are seen in their extreme forms. I make this argument, employing a lesson from Antonio Caso that can help us advance our thinking about the tension between the individual and the collective, hoping to discover an amicable way to relieve it. Following this measure, the attention turns to nonhuman life, future generations, and urban artifacts, striving to give sufficient weight to their significance in the context of urban mobility.

NOTES

1. Hans Jonas, *The imperative of responsibility: In search of an ethics for the technological* age (Chicago: University of Chicago Press, 1984), 8.

2. Robert Kirkman, *The ethics of metropolitan growth: The future of our built environment* (London: Bloomsbury Publishing, 2010).

3. Eugene Hargrove, "Weak anthropocentric intrinsic value," *The Monist* 75 no. 2 (1992): 198.

4. Hargrove, "Weak," 198.

5. Ibid.

6. Ibid., 199.

7. For example, see Lynn Westney, "Intrinsic value and the permanent record: The preservation conundrum," *OCLC Systems & Services: International Digital Library Perspectives* 23, no. 1 (2007): 5–12.

8. Bryan Norton, "Environmental ethics and weak anthropocentrism," *Environmental Ethics* 6, no. 2 (1984): 134.

9. Norton, "Environmental ethics," 134.

10. Ibid.

11. Ben Minteer, "Anthropocentrism," in *Encyclopedia of environmental ethics and philosophy*, eds. Baird Callicott and Robert Frodeman (Farmington Hills, MI: Macmillan Reference USA/Gale Cengage Learning, 2009), 60.

12. Hargrove, "Weak," 206–7.

13. Ibid., 201–2.

14. Charles Gordon, "Automobiles vs. railways for urban transportation," *Electrical Engineering* 50, no. 1 (1931): 5.

15. Jonas, *"Imperative,"* 11.

16. This point suggests that there could be other reasons for supporting such a view, such as obligations toward entities due to well-argued reasons.

17. Samantha Noll *et al.*, "Health justice in the city: Why an intersectional analysis of transportation matters for bioethics," *Essays in Philosophy* 20, no. 2 (2019): 131.

18. See, Emmanuel Levinas, *Totality and infinity: An essay on exteriority* [French original 1961], trans. Alphonso Lingis (Pittsburg: Duquesne University Press, 1969), 213. See Enrique Dussel, *Philosophy of liberation*, trans. Aquila Martinez and Christine Morkovsky (Maryknoll, NY: Orbis Books, 1985), 43.

19. Enrique Dussel, *Philosophy of liberation*, trans. Aquila Martinez and Christine Morkovsky (Maryknoll, NY: Orbis Books, 1985), 43.

20. Robert Figueroa, "Evaluating environmental justice claims," in *Forging environmentalism: Justice, livelihood, and contested environments*, eds. Joan Bauer (Amonk, NY: M.E. Sharpe, 2006), 360–76.

21. This point is not meant to suggest that every historical artifact deserves to be considered as having intrinsic value.

22. Noll *et al.*, "Health justice," 131.

Chapter 7

Love, Respect, and Urban Mobility

Examining mobility technologies within broader socio-material contexts provides a view showing that any transport part, or arrangement of groups of parts, can yield bad or good outcomes under a given set of circumstances. This notion suggests that there are several ways to approach such problems, from increasing multi-modal options to incentivizing particular forms of public and private transit and introducing innovative technologies such as aerial tramways. Keeping tabs on the forward thinkers in transportation and urban planning will yield numerous insights into ways to deal with technical affairs that are above my philosophical paygrade. These people are (or will become) the leaders who can (or will) configure the parts to support moral ordering in a way that help secure worthwhile goals for urban mobility.

When it comes to thinking about how transportation outcomes positively and negatively affect the public, numerous cultural and habitual factors become relevant that will influence outcomes, along with how such results materialize. This notion entails that collectivist societies and individualistic nations will have particular proclivities that expose tensions within the thinking patterns that underpin their respective mobility systems. Even though this reality creates challenging conditions for abstractly discussing transportation as a way to provide insights into continuing efforts to address concrete cases as mentioned in the previous chapters, there are broad notions that we can explore to point us in directions that can lead to acceptable resolutions to mobility concerns within the moral sphere. Bearing the notions above in mind, one such way is to study previous debates that pay attention to the patterns behind such considerations.

For instance, there are several ways to sketch the debates between frameworks that focus on individuals' rights versus society's betterment. In moral philosophy, one of the most well-known examples rests with the incompatibility of deontological (duty-based) approaches and utilitarian theories that side with the larger group. In truncated terms, the tension is as follows: in situations of moral ambiguity, should people act in a manner that should be universalizable, informing them of their duties without consideration of consequences, or should they act to produce the greatest amount of happiness so that the outcomes justify the means?

Although each school of thought can produce cities where people would arguably want to live, both have significant challenges that give reason to hesitate their wide-scale adoption without any reservations. One could argue that these positions do not introduce the idea that we must weigh how the thinking behind moral decision-making shapes who we are as persons. Making choices that pertain to transportation infrastructure and policy involve these notions, meaning that they are not exempt from inquiry. While debates concerning the incongruence above could continue indefinitely, philosophy in and for the streets needs action to mitigate existing and (or prevent) impending harms. For such a task, gaining clarity into these kinds of affairs could benefit from looking at how Antonio Caso framed the tensions between what I interpret as hyper-individualism and forms of unyielding collectivism.

The underlying notion of significance that he reveals is that dealing with this tension can yield opportunities to become more ethical persons. Creating transportation systems through arranging the parts as a form of moral ordering is an opportunity to express values. Which, if we can agree that a nuanced way to think about the difference between morals and ethics as the former dealing exclusively with the status of an action and the latter as the way that we agree to live together based on such notions (i.e., similar to codes of ethics specific to occupations), then defining the ethical parameters of transportation requires that we undertake such actions in concert with other people. This ideal assumes that we all share a common view of at least the ethics, even though we will differ with regard to the grounding and criteria for morality.

For Caso, his view holds that we should aim to do more for others than what you would have them do for you.[1] When contemplating urban mobility, a lesson from Caso's thought is that we must go much farther beyond basic animalistic instincts, which for humankind is our tendency to focus on the economic dimension of a situation.[2] This result is the end that one has in mind if they are not expending the kind of thought required for respecting *persons*, which bring values into view. Caso's outlook provides a way to ground this notion to show why such conditions are inherently problematic:

> A beehive is an individual that reduces each of the individuals within it to the law of the hive. The error of individualism and the error of socialism are very much alike because in their extreme forms both the social theories and philosophical creeds ignore the superior nature of the human being, they ignore the

quality of his [or her] spiritual reality. Individualism and communism lower the dignity of the person. Person and culture, however, are concomitant, for the person requires a society for his [or her] development. Society, in turn, needs the person for its very being.[3]

In the passage above, Caso's take on spirituality involves creating values, which we must do with other people. If we employ his line of thought to help us examine traffic jams, it exhibits that humans as individuals are receiving consideration that puts them on the same level as ants returning to their mounds, suggesting a lack of respect for people as agents. One could argue that such collective behavior reduces us to animals, forcing people into social conditions that reasonable people would not select if one had better options available. Seen in such a manner, humans as parts have their intrinsic value eliminated, reducing them to the status of passive parts, an act that is inherently morally problematic.

This socio-material arrangement is one that each person must *endure*. It creates "endurement," a situation that is not suffering in a tragic sense, but it is an unpleasant experience that we must tolerate because there is no other feasible option. It is long-lasting. Unlike the vulnerable and oppressed peoples who must contend with harms that an immoral arrangement of mobility parts can produce, such as living near a congested, polluting arrangement of vehicles and highways, these people do not suffer on any similar scale. The problem here follows that same pattern, but it manifests in a form that is not oppressive. Yet, it is repressive in the sense that it inhibits and restrains people from engaging in other forms of local travel, which is, in part, a consequence of mono-technical saturation.

Due to this situation, by comparison, we are reminded why vulnerable and marginalized groups are prioritized higher in the process of moral ordering: they are suffering and enduring greater harm than the general public.[4] Although it is not an inherent feature of highway systems, it is a condition that people expect to find in metropolitan areas where personal vehicles are the dominant mode of mobility, as mono-technical saturation. People in countries such as the United States, Mexico, and now China must endure these conditions daily due to the overabundance of these kinds of parts.[5] Most people do not suffer in the same way that vulnerable groups suffer or become exposed to more significant dangers, but they must enter into situations that provide few benefits.

People find ways to make them more tolerable. They listen to music, podcasts, or audiobooks. Although these experiences are intrinsically and instrumentally valuable, they are secondary in the sense that, while commuting, such activities are pursued to help us endure traffic. It would be naive to argue that people who want to enjoy an audiobook search for traffic jams so that they can listen to a memoir or work of fiction while inhaling other people's exhaust fumes. Recalling a point from chapter 3, research shows that most people view commuting as a waste of time.[6]

Although it might seem that I am overselling the significance of traffic jams, here is an analogy that explains why gridlock's place and general, tacit acceptance is worrisome: gridlock is like a bully. Each day, it takes the city's lunch money. It seems routine, but it should not. What kind of a city just gives up on itself, surrendering to traffic jams by accepting them as unavoidable outcomes of design? Considering the enormous role that transportation plays in people's lives and how it shapes the urban sphere, how can any city with such horrible conditions call itself just? Standing up to a bully is a transformative moment. For the life of a city, correcting such situations should hold steady as a pursuit that can support other worthwhile goals.

Applied to a choice concerning the transit system that one should champion for ethical reasons, then, examining the issue through the lens above shows that a well-reasoned motivation needs to engage with this pressing question: How can we discover a way to think about urban mobility that resists totalitarianism for the masses and the kind of extreme individualism that we associate with traffic in several global cities? More importantly, the questions for transportation specialists who use this inquiry take the following shape: What does such a system look like? What conditions must they create for it to exist morally?

We must explore these questions. On the one hand, transport by tyranny is unlikely to draw much support in a decent democracy, which suggests that we cannot force people to ride busses or light rail, especially in regions where such alternatives are significantly underdeveloped. Even though these measures could provide better ecological or egalitarian outcomes, if people do not choose that option without significant coercion, they are likely to lead to protest or creative workarounds. In turn, we know that they are not the conditions that can deliver better outcomes.

On the other hand, considering that people in numerous countries have to "fight" traffic as mentioned earlier, we see how mono-technical saturation is a unique problem for certain publics. It arises out of the need to deliver an arrangement of transportation parts that people can use with maximum independence for the drivers and riders. Without having prescribed one mode of mobility, the catch-all solution for transportation problems, one could argue this exact issue would not be of much concern.

Although advocating for automobility as the primary mode of urban mobility initially sought to facilitate local travel in a highly effective manner, now it fails to achieve that goal in many (but not all) instances, as terms such as traffic jams and gridlock indicate. The use of these terms to describe events that occur while pursuing urban mobility suggests that drivers are not likely to enjoy the ride, and they lack practical alternatives in several such instances. There are probably numerous ways to find a balance between transportation totalitarianism and extreme individualism. Identifying combinations and modes of mobility that can achieve such a balance remains challenging. Choosing a particular way forward that serves this purpose should not emerge from transport loyalty as much as it does to speak to its place in a moral order.

Though we do not know precisely what a moral solution to an extremely complex case would look like in the real world, the situation above shows what we should move away from to create a mobility system that reflects good values that people can develop together. If we entertain the notion that transportation is more than just a means to an end, a way to travel between two points and back again, then it is a dimension of humankind's culture, then we can open our thinking up to a new level of meaning for urban mobility. This idea suggests that we must abandon the pattern of thinking about transportation in ways that focus heavily on economic dimensions, ones that have a history of leading to mono-technical saturation.

The trouble with appealing exclusively to economic elements when considering them in the context of urban mobility, from my take on Caso's position as described prior, is that they block or reduce essential considerations (i.e., such as those that support worthwhile goals) that should carry more weight. We could possibly create much better transportation systems and cities, but making them manifest depends on the willingness and courage to pursue them. If we fail to do so, then we reached peak mobility long ago, and our best is essentially our worst—which includes gridlock on a Saturday, for instance.

The possible worlds for urban mobility that we could create could strive for much better outcomes than appealing to a single, lower-ordered motivation. I would consider this notion to include things that we must produce together that we cannot always measure.[7] Nevertheless, we cannot reasonably dismiss them without significant challenges. These views could include but are not limited to shared respect or love of each other and for the city, caring for one's neighbors and fellow urban dwellers, a sense of community, the betterment of humanity, and the city's aesthetic that a well-moving transportation system can provide.

Bearing the above points in mind, Caso illustrates that they hold much more importance due to the pattern that underpins the moral dimensions of culture as they intersect with personhood:

> Culture is the continuing work of human societies. Culture, however, implies a synthesis of values, and values are constant relationships reflected in thought and action. Further, value or religious experience can never be postponed. In order for [humankind's] social life to take on full meaning, value must be the

predominating influence in human consciousness. . . . Humanity has forgotten love. It no longer thinks in works of charity, but of works of egoism.[8]

While the beginning of the passage above would probably not receive much resistance as an idea concerning the formation of shared values within a given society, the latter could. One can assume that it would provide reason for pause, considering that it deals with ideas such as "religious experience" and "love." However, it is worth mentioning that if such notions appear inherently problematic, it says more about the society or people who are troubled by these ideas that it does about the thoughts behind these notions. Is there any non-arbitrary reason why such ideals are not just as reasonable as any others? "Religious experience" might be too open-ended for a discussion about urban mobility, but "love" might be less difficult.

Still, although one could argue that the academic or professional temper has no patience for the idea of love within philosophical discourse, especially one that is examining urban mobility, there is no good reason to dismiss this notion. This reality says more about those institutions than it does about these ideas. This point aside, if we were to take the above ideas seriously, then what would it mean to create a transportation system out of love for the people? I argue that it would be fantastic, and a step in the right direction for humanity.

However, if such an idea is too absurd for professionals, then perhaps we can begin with the opposite, "hate," to see how it fits in with transportation planning. Would we want a professional to plan for a city's mobility out of hate? Of course not. Nobody (or at least a reasonable person) would want hate as a motivation, and it would not be a hard sell for this view to find purchase. Yet, love might be too intense for some people. Would something in the middle suffice, say impartiality for efficiency? No, those terms will not work because they do not provide us with the requirements to approach transportation systems properly with all of the necessary information to plan, build, or maintain them.

Consider, for example, Andrés Valderrama argues that assuming that professionals already have the information to solve the problem, that their knowledge is superior to other forms of knowledge, is shortsighted.[9] They need a story about how transportation will affect people. Seeking this information in a meaningful way cannot be shallow because such efforts are part of that status quo—which is precisely what we are trying to improve. One way to move away from superficial practices is to bring people into the conversation, which is meaningful because their views can become part of the planning procedure. Such a practice, if it were to meaningfully and thoughtfully include people and their concerns, worries, needs, and hopes for a transportation system that addresses those areas, then it would signal that they would receive *respect*. While asking a transportation professional to love the people might go too far, and hating them is plain wrong, perhaps respecting people is the threshold. Anything above that is positive would be an excellent addition. Anything below that point would solicit criticism.

Along these lines, Caso offers a glimpse into the wisdom that can accompany such an outlook: "We would like that each one be recognized for what he [or she] is: a human person. The idea of person requires respect because it implies respect."[10] In this passage, we must keep in mind that, for Caso, being a person deals with creating values, and respect should hold steady when transitioning from theoretical reasoning to city streets.

While these notions might sound fair, what would a system that makes use of this idea look like? The transportation specialist would need to provide an answer in the form of her design. Any such design would embody the principles expressed above. One way to manifest such ideas comes from the mayor of Bogotá and champion of the world famous, TransMilenio Bus Rapid Transit system, Enrique Peñolosa, who once quipped: "In terms of transport, an advanced city is not one where even the poor use cars, but rather one where even the rich use public transport—or bicycles."[11]

Although his comment sounds somewhat tongue-in-cheek, there is wisdom behind it. That is, the design of the mode of transport, as it is situated alongside social and technological systems, must draw people in so that they want to ride alongside other people. This point implies that, when it comes to looking at how to turn drivers into riders, the sake of efficiency and economic feasibility might not provide the necessary inducement.[12] One way to think about such situations is that it is the *experience* of the bus or the bicycle that must induce drivers to put their cars in park so that they can collect dust. Anything else could be construed as transportation tyranny.

Yet, these notions require that we ask: How can we create better outcomes by managing the parts to deliver a system that can accomplish such a task? This point makes us turn back to Jonas's insightful notion that we need interdisciplinary pooling of resources to get the job done.[13] In turn, moral ordering must be done with the people who will use them in their daily mobility routines and recreation. Although this idea radically challenges the status quo and will require the development of additional measures, if we want to future of urban mobility for the public to not resemble the present, then the inter-pooling of the needed resources must address this aspect meaningfully. Having laid out the benefits of structural ethics and weak anthropocentrism for moral ordering, we must now examine a way forward for dealing with the elements of moral ordering that remain in the sequence mentioned earlier, which include the nonhuman world, future people, and urban artifacts.

When it comes to addressing ecological entities, we cannot do so in a manner that isolates them from anthropocentric issues, ones that extend beyond the immediate horizon. We also need to bear in mind that we remain epistemologically situated so that we can only see our interest for them, despite the tendency to assume that we can "think like a bat" or "think like a mountain." This reality, while being consistent with the tenets of weak anthropocentrism, entails that we cannot merely discount them in an arrogant spout of strong anthropocentrism. The arrangement of the parts that come into play for the above-explored groups will also impact ecosystems and nonhuman animals, and they deserve consideration for their intrinsic value. This notion suggests that they are active parts, but this situation does not change their position when considered in the totality of moral ordering. They can be active parts, but such a designation does not take away from the epistemological challenges that Hargrove raises, as examined in chapter 5.

Nevertheless, considering that we are concerned with outputs centered on benefits for humankind, these points must be considered in tandem to see the interconnected nature of such issues. With such a view in sight, we can focus on the interplay of a transport system's parts and how they affect the moral order. In the section below, I explore these ideas, appealing to the ecologically minded insights required to make sense of such considerations that have effects in the present and will have in the future.

THINKING LIKE A HIGHWAY

Following considerations for the public, we must integrate science into this approach. An Environmental Impact Assessment (EIA) or similar measure could help determine how an action would affect the nonhuman world.[14] Including this aspect into the discussion thus far underscores Jonas's point that such a massive undertaking requires that we pool our resources. Bringing in the relevant experts with the necessary backgrounds speaks to this concern. While they can provide information about how a decision could influence people living today, EIAs can also indicate how a choice might affect future generations.

One could push back against this point, holding that an IEA is too rigid to fit in with an "anti-framework" framework that demands flexibility. That point is fair. However, it fails to consider the reality that science in such a sense does not negotiate. In turn, we must work around it instead of expecting it to work around us. This notion underscores the reason why we need a flexible framework: the world changes and we need to adapt, which could require that we reexamine the parts to see how the world alters the way they interact to produce new outcomes.

In terms of transportation systems, bringing in data from fields such as climate science, forestry, marine ecology, industrial ecology, and public health could identify environmental issues that demand attention to inform decisions that impact human communities now and in the future. This notion suggests that transportation specialists should place our perspective of nature's "interests" ahead of our immediate concerns in select instances. Remembering Hargrove's lesson that we cannot escape being humans with a human's perspective, such a move still counts as being anthropocentric. For such cases, we are acting on our interests and our view of what we think is in the interest of nonhumans. Although this point might sound straightforward, it is rather sophisticated, considering the unique conditions that could pertain to any given consideration.

Consider, for instance, that the nonhuman world cannot be dealt with through employing reductionist measures or one-size-fits-all solutions. Ecosystems will differ vastly, depending on where in the world you are trying to build an urban transport system. Certain parts will only work for particular places, and this notion pushes against the idea that a solution for one locale will work in the next city. Select places might have an issue with urban expansions that displace wildlife, forcing them into metropolitan environments.

On a similar note, road ecology is a relatively under-explored area of environmental and animal ethics that moral ordering can bring into view, a subject that will be more relevant to cities that are dealing with the mono-technical saturation of automobiles than places with plentiful mobility options that do not endanger wildlife and urban animals.[15] This subject could overwhelm the process of moral ordering, but it is one that transportation planners must address. Despite lacking fundamental underpinnings, professionals dealing with such affairs have done so in some instances that emblematize this specific dimension of moral ordering.

For example, Gary Knoll makes this notion evident when examining the history and progress of these kinds of planning measures:

> Early roadkill mitigation techniques prioritized the hardening of the highway, an attempt to make the highway as uninviting to animal mobility as possible. This order is being replaced by the idea of the permeable highway, building an infrastructure that accommodates animal mobility and lessens the impacts of habitat fragmentation. This transformation in highway ideology occurred when members of natural resource agencies, departments of transportation, and local communities came together at early stages of project design. More important, this transition only happened among those who dispensed with a wilderness ideology and viewed the highway environment as a part—as opposed to the antithesis—of the economy of nature.[16]

In the passage above, Knoll shows that the early professionals who were working to find a solution to the entanglement of humans and nonhumans, manifesting horribly as roadkill, employed an approach that failed. Remedying this situation ran against such practices, discovering an approach that examined the arrangement of parts, active and passive, while also bringing in the people who were directly involved in such scenarios. It also supports Jonas's idea that these kinds of problems can directly benefit from the pooling of expertise and backgrounds to gain an interdisciplinary perspective on specific situations. In effect, such outcomes align well with the process of moral ordering in terms of prioritization.

This notion shows that when we think through these kinds of affairs, one can make a case that moral ordering is embodied in the process. It is not that the process was used to make decisions, but analyzing such outcomes can exhibit that engaging in this way of thinking can deliver progress. With this notion in mind, it should seem that it is entirely reasonable to hold that engaging in a process such as moral ordering with an increased degree of flexible and well-ordered reasoning could benefit these kinds of problem-solving measures, especially in cases wherein the experts are venturing into unfamiliar moral territory.

Although the example above only accounts for one particular issue that concerns nonhumans, it shows the kind of situation that one could expect to find when discussing the array of problems that include balancing interests. It also indicates that in cases that involve transportation infrastructure, while inherently anthropocentric, we could change the perception of intrinsic-value being nonhuman active parts, providing them with more respect than seeing them as problematic "parts" that require attention. Even though saving human lives could ground the initial motivation to address such issues, this point does not eliminate the possibility that we can add, arrange, or rearrange the parts so that nonhuman animals are put in advantageous positions so that they are not killed by vehicles. Due to the reality that this approach requires concentrating on outcomes, motivations remain secondary. Yet, we cannot dismiss them because they call us, as moral agents, into question. That is to say, if we begin to question why humans should act for the "interests" of animals, then we return to early papers in animal/environmental ethics, which goes against the histories that those fields established. Still, complex transportation issues that involve enmeshments of humans and nonhumans require approaches that offer a way to balance such "interests" that begin from a place of undeniable anthropocentrism—transportation systems that were designed to move people.

These kinds of issues remind us why weak anthropocentrism is well-suited for addressing such affairs. It provides a way for us to alter the process of moral ordering when a situation such as this one above presents itself in the course of transportation planning. Along with this point, one might wonder why it is that nonhuman life commands a place in the process of moral ordering above future generations of humans. On the surface of the issue, it might seem as if stacking considerations for nonhumans above future generations violates the moral ordering for which I advocate, but this is not the case.

Although future generations of humans come after nonhumans in this arrangement, placing non-existing humans over existing ecosystems is shortsighted. For humans to exist in the future, caring for the environment is a required condition, and humans living today require a non-harmful environment.[17] That is, we can make or retrofit transportation systems to exhibit care for the nonhuman world, preserving its instrumental value while concurrently wanting to preserve its awe-inspiring ability, its intrinsic value. In turn, this approach still supports the version of weak anthropocentrism that I examined above because it keeps humans as the prime focus, but the nonhuman environment is not strictly viewed only for instrumental reasons.

With this point in mind, future generations are the next "part" that "receives" moral consideration in the process of moral ordering. "Part" is qualified because the term requires existing, a quality that future people do not have. Some philosophers would categorize this issue as a version of the "non-identity problem."[18] The issue with considering future generations, according to Jonas, is that it is challenging to say that we have an obligation to future generations because they do not exist.[19,20] However, it seems as if we want to be able to say something meaningful about taking care of the planet for the sake of future generations.

For instance, world leaders such as former president Barack Obama and Pope Francis have urged us to embrace environmental stewardship for this reason.[21] To find an amicable way to address this notion, Jonas argues that we need not look any further than our duties to ourselves: "With this imperative we are, strictly speaking, not responsible to the future human individuals but to the *idea* of [humankind], which is such that it demands the presences of its embodiment in the world . . . thus making it a duty to us who can endanger it."[22] With this passage, Jonas circumvents the problem of non-identity, meaning that we do not even need to have the conversation about obligations toward future generations for their own sake.

Thinking about the magnitude of Jonas's view, the idea of future generations suggests that we are thinking about a topic that is so important that we need not go outside of our own self-interest to give it more weight. One could argue that the idea of future generations has a kind of intrinsic value that is consistent with Hargrove's earlier description. That is to say, thinking about them has instrumental value because it can drive us to make the world a better place to live. However, the idea of perpetuating the existence of the human race goes beyond mere instrumental motivation. It is a good thing in and of itself, suggesting that we can make room for it in moral ordering. This notion is consistent with the tenets of weak anthropocentrism, considering that it accounts for one of humankind's interests. We can square the idea of perpetuating the human race with the other categories within moral ordering, ensuring that our arrangement of parts for the benefit of those groupings does not impact future humans' abilities to become active, intrinsic-value bearing parts.

Lastly, the built environment's intrinsic value deserves attention, even though such cases are highly distinctive. For example, each city has its own unique history, architecture, landmarks, neighborhoods, and social issues. When a problem arises that involves these kinds of elements, one city's solution is not guaranteed to work for another place. Aside from this point, I put the intrinsic value of urban elements (buildings, bridges, infrastructure, etc.) at the bottom of the process of moral ordering because these components frequently change and can be replaced in many instances. In contrast, we cannot replace people, de-extinct species (to their original state), or quickly restore the conditions for genuine human life in the future if we destroy them.

As mentioned in passing earlier, they are passive parts because they do not do anything. Our active participation in urban life is the reason why they exist as parts of the city and, in many instances, parts of transportation systems. Still, what we need now is to secure a place for such entities in the process of moral ordering that gives the degree of respect that their ontological status demands, which is why they, as passive parts, come last. It will be advantageous here to extend the pattern of thinking behind weak anthropocentrism to several areas to cover intrinsic value in cities as they relate to mobility issues. Recalling Hargrove's partiality toward (living and nonliving) nature's beauty and its intrinsic value, such as caves, for instance, most urban artifacts could also have it. If this is the case, then almost anything in the city could have intrinsic value, keeping in mind that urban artifacts could be considered for their aesthetic value like the Mona Lisa or a cave.

According to Hargrovian weak anthropocentric intrinsic value, one could argue that urban artifacts such as elevated trains, trolley lines, and historic boulevards could fit under this description. Perhaps people will not agree that a freeway exit is an awe-inspiring work of art, but expert designers and engineers could easily disagree. The same notion applies to any part of transportation infrastructure that connects the city. They might not have much aesthetic appeal to residents, but such technologies have the possibility of being aesthetically appealing to a point that transcends mere appreciation.

What is more, if having enough beauty to transcend mere aesthetic value can serve as the basis for intrinsic-value arguments, we should also inquire about similar grounds such as historical and/or cultural significance. Suppose we can agree that humans, nonhumans, and artifacts deserve consideration based on the standards of weak anthropocentric intrinsic value. In that case, concerns should rest on how we prioritize them, bearing in mind that we cannot always preserve everything that has such values. Denying this realism implicitly (and erroneously) could suggest that sidewalks deserve the same consideration as people.

In turn, such a notion could motivate us to ask: "Do all urban artifacts warrant consideration?" The short answer is "yes," but the degree of such consideration, along with how strongly that we should view it, exhibits that there are limits to them. In some instances, such as with the intrinsic value of a speed limit sign, they could be incredibly short-lived. This notion rests on the idea that, as Hargrove argues about ecological expertise, we have to rely on well-informed assessments to determine if an artifact has such value. Although in trivial cases such as removing a speed limit sign, there is little reason to worry that someone would protest through appealing to its intrinsic value. However, there are serious cases that involve meaningful topics that include quality-of-life issues, such as numerous affairs that bring urban mobility into view[23]

For example, consider highway removal projects. Such undertakings show that getting rid of a highway that has proven to put residents at a disadvantage clearly indicates that such a structure lacks instrumental value. Removing it would serve the people in some capacity.[24] On the contrary, one could argue the highway's historical significance gives it intrinsic value that deserves *some* consideration. This example shows that we have to debate who gets to count as being qualified to make assessments about intrinsic value related to existing highway projects that currently serve some city dwellers and commuters. With this point in mind, we cannot easily dismiss such concerns, meaning they deserve some degree of meaningful consideration. Investigating these aspects provides an opportunity to explore the process mentioned earlier that can shift our thinking on mobility matters, providing us with a way to deal with several groups that deserve moral consideration, which is moral ordering.

While this configuration addresses the problem of moral prioritization, it does not assume that such an arrangement is without exceptions. Similar to SOP-1 and SOP-2, conflicts will arise when formulating a response wherein it appears that moral prioritization fails. Consider, for instance, if a tract of land is set to be developed for affordable housing for marginalized people. However, an EIA determines that doing so would devastate a nearby wetland that provides numerous ecosystem services that are essential for the community. If we employ moral ordering as described above, we will lose the wetland, flood control, water purification, and wastewater treatment. It would harm several nonhuman species that the locals revere.

It seems that acting for nonhumans in this case, instead of acting for marginalized people, would be a well-supported endeavor, but it would be one that would appear to go against moral ordering as described above. While it might seem that impeding construction would fail to comply with moral ordering, preserving the wetland would actually be of greater importance, considering that marginalized groups and the non-marginalized public would tremendously benefit, along with nonhumans and future generations. In turn, such efforts would not go against moral ordering. These kinds of circumstances justify giving moral ordering a flexible nature.

However, the significant notion that must remain in our minds when attending to such matters is that the call above would be a tough one to make. This decision could be met with protest unless an alternative option was given more consideration so that it outweighed the decision to spare the wetland. Seen in this way, the decision to spare the wetland would benefit the people for whom it would serve, which includes the marginalized group, but such an effort only *begins* there. It would be paramount to continue efforts to find an arrangement of parts that was morally optimal, which must explicitly entail meaningful participation from group members (unless they were to protest their involvement). This kind of arrangement would need to deliver the outcome that sat well with all parties, primarily the people who come first in the moral order. This effect would need to be the case unless such a compelling reason would warrant otherwise, which requires examination on a case-by-case basis. Under no circumstances can an answer simply not offer powerfully compelling reasons.

When dealing with multi-tiered conflicts, solutions that appease all parties will be challenging. For example, arguing that acting in humankind's environmental interests while simultaneously acting on behalf of nonhumans' intrinsic value to develop high-rise affordable eco-housing that includes tearing down a culturally historic building could be overwhelming. The historical significance of such a structure for some people might be a more substantial interest than preserving an ecosystem. Settling such cases would not come easily, but these conditions do not mean that moral ordering must collapse into relativism.

On the contrary, these conditions suggest that we must face hard questions, but not impossible ones, about moral rightness in such instances. Simply because we might not discover a moral solution does not entail that one does not exist or that similar cases are doomed. Nevertheless, these challenges come with problems such as climate change and, in general, learning to live on a complex biosphere.

Despite the challenges that exist in a world that lacks inherent uniformity and dealing with cities that lack entirely consistent characteristics, the direction shown thus far illustrates that we can break urban mobility problems down in a manner that makes them manageable. Through employing a mereologically inspired approach that provides a way to see how transportation systems as wholes and their parts, along with the relationships between them, there is hope that we can deal with them in a manner that brings the moral dimensions of such affairs to the forefront of our thinking. With this point in mind, we can employ these lessons to pave a way forward that can help us realize that there are much larger ambitions and goals that we can pursue to create worthwhile urban mobility. Chapter 8 begins exploring these ideas.

NOTES

1. Jorge Gracia, "Antonio Caso," in *Latin American philosophy in the twentieth century*, ed. Jorge Gracia (Amherst, NY: Prometheus Books, 1986), 41–2.

2. Gracia, "Caso," 42.

3. Antonio Caso, "The human person and the totalitarian state," in *Latin American philosophy in the twentieth century*, ed. Jorge Gracia (Amherst, NY: Prometheus Books, 1986), 44.

4. The fact that I feel that I must argue this point so strongly is not so much a reflection on the position itself or my view of it, but it shows that the modern temper of some kinds of academic philosophy have created the conditions wherein I feel compelled to ground arguments for why we should help people who are suffering badly should have an increased sense of urgency over people who are not suffering to the same degree should be an alarm.

5. Trevor Reed, *INRIX global traffic scorecard*, 2019. https://inrix.com/scorecard/.

6. Patrick Singleton, "How useful is travel-based multitasking? Evidence from commuters in Portland, Oregon," *Transportation Research Record* 2672, no. 50 (2018): 14–16.

7. Of course, there are elements that we can measure such a public-health outcomes and emissions that are extremely significant.

8. Caso, "Human person," 45.

9. Andrés Felipe Valderrama Pineda, "What can engineering systems teach us about social (In)justices? The case of public transportation systems," in *Engineering education for social justice*, ed. Juan Lucena (Dordrecht: Springer, 2013), 215.

10. Caso, "Human person" 44.

11. Enrique Peñalosa, "Why buses represent democracy in action," *TedCity2.0*, September 2013.

12. In the transport planning literature, such discussions are housed in the context of "captive" versus "choice" riders, indicating that there are larger, deeply seated elements to consider. Although exploring the pitting of these terms against one another, for more information, see Jinhua Zhao *et al.*, "Customer loyalty differences between captive and choice transit riders," *Transportation Research Record* 2415, no. 1 (2014): 80.

13. Hans Jonas, *The imperative of responsibility: In search of an ethics for the technological age* (Chicago: University of Chicago press, 1984), 189.

14. US Environmental Protection Agency, "Environmental impact assessment," http://www.epa.ie/monitoringassessment/assessment/eia/.

15. This point is not intended to suggest that there is an entire gap in the literature on this topic in these subfields. For example, a recent paper that explores the consumption of roadkill, see Cheryl Abbate, "Save the meat for cats: Why it's wrong to eat roadkill," *Journal of Agricultural and Environmental Ethics* 32, no. 1 (2019): 165–82.

16. Gary Kroll, "An environmental history of roadkill: Road ecology and the making of the permeable highway," *Environmental History* 20, no. 1 (2015): 4.

17. Robin Attfield, "Beyond anthropocentrism," *Royal Institute of Philosophy Supplements* 69 (2011): 46.

18. For a recent work that addresses the non-identity problem in great detail, see David Boonin, *The non-identity problem and the ethics of future people* (New York: Oxford University Press, 2014).

19. Hans Jonas, *The imperative of responsibility: In search of an ethics for the technological age* (Chicago: University of Chicago press, 1984), 38.

20. This claim is also present in Richard Routley *et al.*, "Nuclear energy and obligations to the future," *Inquiry* 21, no. 1–4 (1978): 143.

21. Barack Obama, "Weekly address: Protecting our planet for future generations," https://obamawhitehouse.archives.gov/the-press-office/2015/10/24/weekly-address-protecting-our-planet-future-generations.

22. Jonas, *Imperative*, 43. The original term that Jonas uses in this passage is "Man." I updated it with "humankind" in brackets to indicate the replacement.

23. It is worth mentioning that in some cities, having no dumpsters is a serious quality-of-life issue.

24. City of Seattle, "Case studies in urban freeway removal," http://www.seattle.gov/transportation/docs/ump/06%20SEATTLE%20Case%20studies%20in%20urban%20freeway%20removal.pdf.

Chapter 8

Moving, Thinking, and Cooperating

People often claim to love their cities, but the question remains to be asked and answered: Do they love them back? People participate in making cities great through their culture, labor, and business, supporting them through these and similar measures. In an imaginative sense, we can say that such actions are a way to show their city love. Yet, do the ways that we have structured cities make people feel safe, welcome, and—loved? This point suggests that there is a kind of relationship going on between residents and their places. In turn, one can argue that individuals are subject to influence from their socio-material environments, at least to some degree. This notion goes along with the claim that technological parts can play roles in the morality of urban living.

The process of moral ordering zeroes in on this idea and—as a technology itself—encourages us to focus on shaping the outcomes that we desire. That is, recalling from several chapters ago, it is a theoretical device that helps frame situations in ways that move us to work on mitigatory efforts or solutions, giving attention to several groups of stakeholders in what aims to resolve the issue of moral prioritization regarding transportation systems as wholes.

Considering the significant impacts that cities and regions have on people's lives, taking the above notion seriously requires examining the reciprocity at play between residents and the places they call home. Specifically, looking back at the previous chapters to illustrate a picture showing the magnitude of how transportation systems affect people, we need to pay close attention to this dimension. We have seen how transportation systems play significant moral roles, affecting every group, as examined earlier through moral ordering.

One of the more prominent areas that have made progress when it comes to people who are caught up in such arrangements from which we can gain inspiration is the traditions that Henri Lefebvre established on the right to the city. For instance, researchers, such as David Harvey, who draws from his work, argue that urban residents have a right to shape the city, focusing on the forces that control urban places.[1] Along with the insights that stem from this position, one can argue that we still need specific direction for moving the conversation forward in terms of real-world measures, ones with tangible effects. Connecting such ideas to the concepts that will bring them into reality should hold steady for future research. This point does not discount the significance of analyzation, but it seeks to balance theoretical dimensions with the necessary feet in the street to get the job done.

To think about people shaping transportation systems implies that they will help reshape the city, which will provide residents with the necessary means to undertake pursuits that will determine several dimensions of their lives. They are not merely shaping the city. They are shaping their health, mental life, time with family and friends, happiness, and how far they can advance socioeconomically.[2] Adding and/or rearranging the parts of a transportation system that can save them several hours per week could provide residents with needed time. They could engage in the sort of self-determined practices that would facilitate pulling themselves up by their bootstraps or taking off their boots to relax after a challenging week.[3] Perhaps the hope is that one day that they can replace their work boots with Oxford shoes—instead of dreaming that their children will be the ones to have a better life that includes comfortable footwear at the office instead of the factory, warehouse, or field.

The point here is that, through acquiring extra time, other goals become feasible. The aspect worth highlighting is that through employing such an ability, residents gain a way to influence the sociotechnical means that will define the contours of their urban experience. This idea suggests they can, to a degree, obtain the kind of influence that leans toward redesigning their city and their lives authentically. While these points sound uncomplicated, having these conditions materialize would include several steps that possibly only professional planners can identify, along with their associated challenges.

In turn, this reason supports the idea that members of this profession are tasked with more than making the trains run on time. They have the ability to help people secure the necessary footing that can launch them out of poverty and/or achieve a better standard of living. While there are many reasons people can be vulnerable or marginalized, which puts them in precarious positions that most residents do not face, at least in many cities, the specificity of such conditions will change. This idea reminds us that such circumstances will differ from one city to the next one, bringing elements such as unique histories, atrocities, and victories to conversations about urban mobility that will only make sense in their respective contexts.

Such a notion does not suggest a kind of "urban relativism," but it does support the idea that, even though issues in transportation share characteristics, they differ. This point deserves restating because attending to such matters successfully bolsters other areas of concern that hinge upon it: housing, food security, and recreation—just to name a few. As indicated previously, transportation is what literally connects people to these necessities. Developing a plan of moral ordering could sound challenging, but creating measures that achieve results that prioritize them within a plan to move the public, I imagine, is no easy feat.

Although pursuing such measures might sound straightforward, the reality of the situation is not. One can argue that the parts of transportation systems that need to change or be removed are deeply rooted in our cities and in the minds of the people who control them, along with differing degrees of social acceptance and embrace. For the transportation specialists and municipal officials who undertake these jobs to overcome such seemingly insurmountable goals, meeting them with success could warrant moral praise in select cases. This idea also goes against the notion that people engaged in these occupations are simply doing their jobs in all such instances. On the contrary, they are attending to the affairs of human suffering, flourishing, and thriving. The arrangement of the parts matter and underestimating their political dimensions hinders a comprehensive view of how they can impact people.

Although the cases explored in earlier chapters illustrate how transportation technologies can harm minority groups, such accounts do not provide insights into the extent that individuals can be harmed from such arrangements. Despite knowing about transportation injustices, I cannot adequately express how transportation infrastructure affects people in situations other than mine. Still, we have no good reason to doubt people whose lives are excessively burdened and revolve around transportation systems when they do or can tell us about what they face and battle when merely going to work, get groceries, or run errands.

Consider, for instance, *Adela's Journey*, a short documentary video, that illustrates how Adela, an immigrant living in Portland, Oregon, works tirelessly to survive.[4] She deals with a transportation system that is all but hostile toward her existence. Due to the arrangement of transportation parts, most elements of her life revolve around getting to and from part-time jobs. The arrangement of parts, including but not limited to the bus, the bus drivers, the limited schedule, and the bus stops, played a role in her not getting better employment, feeling unsafe, puts economic strains on her life, and limits her time with family.

In the video, she explains to the filmmaker many of the logistics that cause her burdens to increase. One could argue if her reality were evident to planners who cared, then it could be the case that filmmakers could highlight the tremendous job they did to lessen her burden. Due to situations like Adela's struggles, grassroots groups such as OPAL Environmental Justice Oregon have undertaken several campaigns to bring attention and gather support against such conditions, illustrating what self-determined leadership looks like.[5]

Although the above example is merely anecdotal, one could argue that it is emblematic of numerous people's conditions worldwide. Each instance has unique characteristics that speak to geographic differences. These points aside, the example above suffers from being a highly insufficient sample size. Another way to appease the modern philosophical temper is to provide a fictitious example, a "thought experiment," that offers a different approach from the above narrative. To gain a concrete but brief illustration of the importance of planners' work, imagine the two versions of the same situation described below.

In the first case, a single mother struggling to improve her life and her child's future is on her way to a job interview during the summer. She spent the previous days researching the position and roleplaying question-and-answer sessions with a friend. She put together her best outfit, one that increases her confidence. Secured childcare. Now, she is standing on a sidewalk by a single bus stop sign without a protective covering, which is typical. She waits for the bus. It is running late. Already nervous, panic sets in. Due to the unforgiving sun, beads of sweat turn into streams. To calm herself, she thinks of the extra blouse and baby wipes, neatly tucked away in her purse in case of a last-minute emergency.

Finally, the bus arrives, about ten minutes behind schedule. Luckily, she planned to catch an earlier bus, anticipating such events because she takes this line frequently. She knows to expect the unexpected. Once onboard, she carefully selects her seat. Previous incidents taught her to check for unsuspecting liquids, foods, or items that previous travelers left behind. Arriving at her stop, she thanks the driver, even though he is silent upon her exit. His impartiality does bother her, but she is unwavering in her efforts to express gratitude. Her parents taught her always to thank the driver. She knows that the operators frequently change on this route, and some of them might not see the point of getting to know the passengers.

Due to her foresight, she arrives early. Ducking into a nearby diner to visit the washroom, she applies fresh deodorant and prepares herself to feel confident, again. She sits in a booth, rehydrating. She is ready. When the interview comes, all aces. Considering all that she has been through today, the interview itself was the last step in an almost impossible obstacle course.

Yet, we cannot see that before she even applied for the position, numerous social, political, and economic hurdles were in her way—systemic traps. While undoing the other realities that these deeply entrenched conditions help perpetuate, let us reimagine the above scenario—after it has received attention from transportation specialists who took on the task of attending to the actualities that a previous job seeker had to endure.

A single mother prepares for an interview in the same manner described above. However, in this case, she caught the last bus because it never runs late. She was able to relax on a bench under a shaded roof. The bus arrives on time. As she boards, the smiling driver, whom she knows by name, greets her and wishes her luck. He knows the woman's backstory. He is rooting for her. Once aboard the clean and Wi-Fi-enabled bus, she watches videos on confident body language, further preparing herself until the last minute. Considering that the bus is always sanitary, she chooses her favorite seat in the middle. The bus arrives on time for her interview. In the end, the job is hers.

Comparing these cases, the former shows that along with the troubles that accompany her journey, a knowledge component remains in play. Her survival skills reveal it when we look at them closely. For instance, inductively, she knows to take an earlier bus. This situation does not arise out of paranoia. It comes from experience, and the same can be said about the dangers of seat selection and the turnover rate of new drivers. Carrying a change of shirt and hygiene wipes, unfortunately, are elements that enter one's mind when there are substantial motivating factors to consider. These things change, depending on the destination and the purpose of the trip.

In contrast, the latter example shows a passenger whose mind, while not at ease due to the stress of an interview, is not under pressure from having to contend with a sketchy public-transit affair. One could speculate the riding the bus in this instance could have reduced her anxiety because she did not have to worry about driving, traffic, rude drivers, or parking. While these hypotheticals are only intended to demonstrate two polarities of transport service, they show that there can be vast differences in the realities of urban mobility. This point suggests that when it comes to planning with moral ordering as a guide, the latter example deserves study so that bus services can embody the necessary degree of respect that individuals deserve, balancing such needs against considerations for the many.

Even lacking the technical expertise that is required to explain and support such a position, it should not be challenging to argue that developing a bus system that favors the second case should not be a head-scratcher. It is a case of prioritization in the grand scheme of a whole transportation system wherein professionals can arrange the parts to deliver the outcomes that can make daily travels bearable. This notion would be incredibly convincing if the city in question already had a dominant transportation system that had produced mono-technical saturation. In turn, re-creating or developing a bus system as described in the preferred instance would benefit passengers and the owners of private vehicles who share the roadways. These are the kind of outcomes that would align with moral ordering, albeit imperfectly, meaning that better outcomes should remain targets. Progress always, perfection hardly.

Although the above thought experiment is merely illustrative, one could argue that the woman's experience in the first case makes her stronger. Due to the many obstacles that she had to overcome, she is a better and tougher individual because they stood in her way. One could even go as far as to claim that her actions prepared her to do better during the job interview. What is more, people *need* these hindrances to help make them stronger. She should be grateful for them.

Bearing these points in mind, we should apply the same pattern of thinking to other forms of transportation. For instance, when highways become riddled with potholes or the painted lines that separate traffic lanes fade, not fixing them will force drivers to become mentally sharper and quicker, learning to deal with the conditions that have emerged. When roadways become too crowded, we should not look for ways to decrease the number of cars on the road, but we should force drivers to become more competitive, outmaneuvering their fellow drivers. If they learn to be competitive on the road, then perhaps this way of thinking will transfer to their other areas of life, helping them get ahead in the workplace. Not attending to these mobility matters would make them become the best versions of themselves. If we are not going to improve public transit to help people become better, then we should extend this courtesy to all modes of urban transportation.

Of course, this example is ridiculous. However, exploring it shows that the former instance is also irrational. If deciding that transportation is not an appropriate venue for testing people's mental endurance and character in one capacity, then it is only fair not to do it for people who have fewer advantages, either at the starting line of life or as they trudge along the happy roads of their destinies coproduced with others. Recalling Caso's point about respect, making both forms of transportation more enjoyable serves as a suitable option. Going one step further to fully include this kind of position, a person doing more for another person than he or she would do for oneself could include adding solar-powered fans to the bus stop canopy and installing Wi-Fi in vehicles (which some buses now have). Aligning this kind of thinking with moral ordering, professionals can attend to essential matters such as protection from the elements before tackling a monstrous highway project exhibits that they want to help the people who need it the most.

However, if the pothole situation on a highway leads to fatal or tragic events, then an explanation holding that we should prioritize such repairs would be sound, meaning that it would not require a cogent moral defense.

This point does not entail that we should abandon upgrades to bus stops. Alternatively, it could also show that workers should expeditiously install the canopies to begin repairs to roadways in need. Whichever the case, it does show that attention could momentarily turn to repair damaged roadways for safety, a move that would show respect for the masses that have to deal with those conditions as part of their daily routine.[6] Although such cases are commonplace in cities where mono-technical saturation is the norm, dealing with them on a practical level will require individualized studies due to the differing nature of transportation problems.[7]

This account exhibits the *suggestive* nature of moral ordering, meaning that its lack of rigid structure is a strength. Being adaptable is an asset, despite lacking an unwavering appeal to universality. Along with this notion, it is worth restating that specialists can make a case for switching the moral order, and even attending to such a troubling matter shows that it is receiving attention. The one caveat that must remain in view is that the motivation to change the order must be scrutinized. Although this point might sound direct, exploring it reveals a tension that requires confrontation to see the complexity and inherent troubles that can arise in some instances.

For example, referring back to the idea of residents having a right to the city, which extends into urban mobility, what are the benefits and challenges to this view? For the most part, investigating such affairs will take us into mostly uncharted territory, pushing against deeply ingrained beliefs and established practices. It would seem rather cavalier to simply pursue a desire to change it without providing a survey of the theoretical ground that requires covering. That is, essentially, addressing the example above exposes the tension between non-experts and specialists, an issue that correlates to a pattern of argumentation that extends into numerous domains. To thoroughly explore this concern and its broader implications for society, the section below undertakes this task. It reveals that while confronting the tension will require considerable attention, the benefit of examining the conditions that pertain to it can yield the beginnings of a way forward, one that could deliver a future that does not resemble the present for urban mobility.

EXPLORING THE TENSION BETWEEN NON-EXPERTS AND TRANSPORTATION SPECIALISTS

The tension between non-experts and specialists has an extensive history of debate in the public sphere, which remains in play today. George William Russell's insights emblematize one side of this debate, encapsulating its pith:

> The desire to mess things and interfere is strong in the official oversoul, and at last a new generation of officials arise, more intellectual, better instructed, real experts in their way, and they begin the whole business of interference over again. Our theory, which we have often put forward, is that experts ought to be on tap and not on top.[8]

Russell's insights are over 100 years old. Yet, they remain incredibly relevant for crafting new ways of thinking about urban mobility. We need mitigatory measures that bring the sentiments behind Russell's assertion into view. In addition, we also need to establish limits for them due to the dangerous nature of transportation affairs and to know full well that any sort of immediate "revolution" in mobility thinking could be disastrous if not ushered in cautiously. These efforts involve developing a way that can *at least* work toward relieving or transforming the tension between transport professionals and the people they serve.

Within the passage above, Russell is not dismissing the need for experts. Acknowledging that they are necessary implies that they ought to serve those who require their insights to facilitate goal achievement. This attitude is well suited for people who actually want to undertake enterprises that would benefit from having expertise as a form of consultancy. Seen in such a manner, the hope that we can arrive at a place of mutual understanding that relieves the tension is now the topic we need to address.

In previous chapters that dealt with conflicting interests between stakeholders, we can say that these matters signal a tension between non-experts and transportation specialists. They remain constant when attending to these affairs. Rather than pitting specialists against non-experts, the productive way to deal with tension is to dissolve it. By showing a fleshed-out understanding of the issue, there is no tension, only a misunderstood situation. That is, why would a person go through extensive training to understand how the parts of a transportation system fit and function together, only to depend on the advice from people who lack such training?

Second, why would people, such as the woman in the previous thought experiment, want to complicate their schedules by giving information to people who spent years studying the nature of such problems? Further, considering that she spends her time advancing efforts for a better life, when would she have the time to engage in the required dialogue? Although she could provide insights into how a specific subset of arranged parts creates nerve-racking circumstances, there is no guarantee that she has the requisite knowledge of how the other parts create the dreaded the situations she must endure.

Viewing the above questions in tandem gives us reasons to see why some planners do not thoroughly rely on members of the public, along with the reasons why they do not often demand to be involved in such decisions. These reasons could include the following: most people lack technical training; gathering input demands too much time from residents; people's views are partial at best. With these notions in mind, it is no wonder, then, why participatory planning approaches lack a history of sustained and flourishing momentum alongside mainstream efforts, even though such efforts are slowly advancing.[9] Aside from these points, there is still a need to ensure that the people who are affected by transportation can meaningfully weigh in, which can help us work toward dissolving the tension—if we maintain that people should have that ability. Yet, along with notions such as political recognition, there is a much more robust reason why we should want to radically attack the deeply seated beliefs holding that such a practice should remain untouched that syncs with the right to the city, which I explore below.

Getting at the bottom of this notion shows that the significant tension is not between the residents and the transportation experts per se. Both parties are operating within a set of social and political parameters that they did not define. Speaking to urban travelers, one can assume that it is customary for people who are raised in a given city not to question their transportation system because it has always been in the background. Due to its familiarity, one can argue that people simply learn to navigate the cityscape if they love it, hate it, or remain impartial.

When it comes to the experts, they are entering into a realm that was there long before they were born, meaning that they have to contend with existing urban structures and infrastructures, the nonhuman world, social, political, economic, and religious influences. Even though they could very well know how to solve particular problems in terms of the basic elements that pertain to the process of moving people, they cannot move the minds of the people who control the means for people to move about the city. Such situations of course will vary when surveying cities, but it seems challenging to realistically hold that planners and engineers can advance their efforts without considering the items mentioned above.

Considering these points, it might appear evident why residents and experts lack congruence when it comes to how they view the world of urban mobility. Yet, they can create compatible outlooks that are productive and complementary by fostering the conditions for thinking about urban mobility in a manner that seeks to put distance between the above-mentioned forces that helped make the circumstances that challenge them. This idea implies that they could deal with the reality that transportation systems were brought into this world in a piecemeal fashion in the same manner.

That is, cities dealing with mono-technical saturation of automobiles did not experience such situations overnight. In turn, thinking them through, perhaps one part at a time, means that transportation specialists must fundamentally think for themselves in terms of the extraordinary efforts that are required to make life easier for travelers. This thinking could include casting aside preconceived notions that they inherited from their profession. Recent advancements regarding multi-modal planning and notions such as "complete streets" that encourage such actions indicate that some kinds of thinking are already geared in supportive directions.[10] Along with such aspects, they must also attend to the matter of people not being able to think for themselves about transportation due to its familiarity and assumed acceptance through passive participation as described above.

Due to the significant nature of such a challenge, transportation professionals who can succeed in such pursuits go far beyond their occupational requirements. Such outcomes are supererogatory, supporting the idea that they deserve moral praise when they find success. They begin by helping people use their own reasoning when it comes to such matters. Considering that we are dealing with affairs of the city, referring to this goal as urban enlightenment makes sense.

We can find the roots of this issue in Immanuel Kant's essay, "Answering the Question: What Is Enlightenment?"[11] Although he was talking about topics pertaining to *The Enlightenment*, we can employ the pattern of thought behind his sentiment to guide people into thinking for themselves on urban affairs with the requisite expertise that could facilitate such an enterprise.[12] For instance, Kant held that we frequently do not think for ourselves because there are people who can think for us.[13] He called for people to gain the courage to think for themselves, to use their own reasoning. It is absolutely not the case that I think that urban dwellers cannot use their own reasoning or think for themselves on such matters or any other issue.

What is more, I do not believe that a lack of courage is what is holding people back in such cases, as described in the latter part of the thought experiment above. The other elements such as time constraints, distrust of public servants and the city, language barriers, and unfamiliarity of planning could impede their interest in participation. Despite such aspects, one could make a cogent case that having them involved in the process could provide direct access to some of the measures that would benefit them and other residents. One way to elucidate the critical nature of such a notion is to ask, what does it take for urban residents to have the ability to use their own reasoning in terms of the "good life" in the city—when it comes to the very issues that shape their experiences and the quality of their lives such as transportation?

Before we can proceed further with this inquiry, another question demands answering: "What is this business of 'urban enlightenment'?" It requires that one has gained superior knowledge and understanding of the city. Transportation is one area of required knowledge. Other areas could include but are not limited to zoning, law, history, architecture, infrastructure, business, and social customs and traditions. While this list is by no means exhaustive, it indicates the kind of elements that should be on it. Working toward urban enlightenment also means knowing how these elements fit and work together to shape a city; why an individual has a particular view and unique experiences, which includes bringing one's situatedness into perspective, along with why other people have their distinct positions on the city that can differ drastically. This point does not imply that they can account for another's subjectivity, but they can know about the conditions that shape it.

These matters of experience—coupled with the reality that they differ for different people—suggest that issues of morality are involved, such as the distribution of harms and benefits, which open up the possibility to examine the preconditions and conditions that produce such outcomes. It is also fitting that we bring into view the idea that people should be able to influence such outcomes. Working toward urban enlightenment, then, not only requires that people understand the composition of a city, but they also comprehend that its composition, as an arrangement of parts, yields different moral outcomes for various groups of people—along with the full range of categories from the process of moral ordering. To bring these notions into the picture, they must know how "the city" affects its vulnerable and marginalized residents, the public, nonhuman life, that it could harm people in the future, and urban artifacts.

This notion, however, also applies to professionals. That is, for them to acquire the knowledge that is required to understand transportation and the city *more fully*, they should strive for urban enlightenment by seeking to learn how the parts of a transportation system affect each category in the sequence of moral ordering as it appears, along with available perspectives that can enhance their knowledge of transportation systems, as parts and as wholes. Although many of their work's demands situate them to understand such elements, the history of lacking perspectives of marginalized and/or vulnerable populations suggests that there is reason to strengthen their view in some instances.

I can imagine some planners challenging this claim, holding that standard practices call for gathering input from residents. If this is the case, are they willing to defend the view that residents decided to harm their children via pollution during a meeting on urban mobility? Were they willing to have their neighborhoods destroyed for a highway? Did they decide that bus stops in the desert did not require protective coverings? I would like to see the minutes from those town hall meetings.

The point here is not to vilify the planning profession. Today's transportation professionals should not have to pay for the sins of their fathers. However, dealing with the cleanup of the past should not waver while planning for the future. This work includes learning from mistakes and taking steps that meaningfully entail inclusivity and the knowledge that comes with it. This aspect counts as the buy-in that professionals could need to take urban enlightenment seriously.

Without such access, they would lack an encompassing perspective on the urban condition as it pertains to mobility in the city. If we consider the idea that such a view is necessary to get the job done morally, then making room for gaining this exclusive form of knowledge should hold steady, alongside other typical elements such as codes, policies, and technical measures germane to their professions. For the professionals who have already taken the requisite steps to secure this knowledge, one could say they are already on the path to urban enlightenment.

Bearing the above points in mind, recall from chapter four that a city is in a perpetual state of change, meaning that knowledge of its state will also be incomplete in most instances. This notion suggests that it might not be a good idea to think about urban enlightenment as a state that one can reach. Yet, it is a journey that one must undertake continually, depending on the degree and rate of change within a given city. In turn, for a person to say that they attained "urban enlightenment" would be suspect (and arrogant). Still, we do have reason to believe that there are people such as planners who have ample knowledge that they could share with people who want it, even though it would only provide partial knowledge. In any event, sharing it through such measures could help people understand how a city fits together and the good or bad outcomes that it produces, shaping life for urban dwellers.

For instance, it is not challenging to think that people who experience and reflect on the array of realities that transportation systems produce are not keenly aware of them, along with how such conditions affect other areas of urban life. With this point in mind, the challenge here, for professionals, is to bring those reasonings to surface so that "the city" can improve its view of how a transportation system produces such outcomes. While municipal representatives can have some knowledge of such situations, gaining more first-hand accounts shows respect for the people who have endured such experiences, and they could bolster their existing knowledge as it pertains to relevant affairs.

As shown in previous chapters, transportation affects all lives, human and nonhuman. Seeing how and understanding why the parts exist in their particular states, along with moral ordering for transportation, which includes a particular city's backstory, urban dwellers can make sense of their cities. It is arguably safe to assume in many cases that urban dwellers already possess most of the knowledge that is needed to meaningfully weigh in on their transportation systems. Although the specifics of such cases will vary across a broad spectrum of possibilities, becoming aware, angry, distrustful, and/or hopeful of a given situation could spark a desire for stakeholder-ship wherein the right to the city in a particular sense, such as mobility, is actualized.

Due to having these insights, residents could gain an enhanced understanding of their transportation situation—one that reveals the power dynamics and conflicting moral issues as examined through moral ordering—along with the moral complexities associated with such arrangements. This idea entails that urban dwellers do not simply state their obvious preferences for what they want to gain from a transportation system without any external considerations, but they would examine their position as it relates to other positions in moral ordering, either embracing or coming to grips with the reality of the totality of mobility affairs in their city. This point is not meant to argue that people should not care about how transportation affairs affect them, but that they should not disregard how their mobility choices impacts others in the moral order. Taking a highly disproportionate position that narrowly favors one's self-interest, considering that urban mobility systems are technologies that we share with other people could, in some instances, be equivalent to basic animal instincts, expressed as the notion that we receive from Caso: economics.

That is, it hardly sounds reasonable that each person appealing to his/her/nonbinary shallow self-interest who lives in a city is striving for urban enlightenment. Could one even say that a place where people only care about their own narrowly defined self-interest is a city that social beings want to call home? Does not the fact that the people have designated or accepted through participation, the city as a municipality—as a locus of local decision-making—indicate that cooperation with other people is present at least on the most fundamental level? One could argue that, by choosing to remain in the city, residents have tacitly agreed to abnegate some power to the city's professionals to make the trains run on time. In a loose sense, then, such limited engagement counts as a weird kind of "non-acting" that "happens" in concert with others.

The urban-enlightened dimension comes into play when people seek to not only freely turn over *select* decision-making abilities that they recognize require expertise that they lack, but they robustly participate in urban life through providing the informed support that contributes to the city in a manner that is reflective of their place in moral ordering. For example, ordinary residents do not need to "use their own reasoning" when it comes to mixtures of concrete for designing bridges, but they should weigh in if such bridges would restrict their movement. Although such an action requires acknowledging the distinction between giving up an ability and recognizing that one furthers their well-being and overall position if they undertake the necessary abnegation. Still, there are limits to what counts as acceptable, signaling that clear parameters should exist within moral ordering.

This notion entails that people who are either vulnerable or have been marginalized are not expected to compromise themselves for others who are in a better position. Such a proposition hardly sounds fair. Yet, when people who have the ability to hang up the keys to their vehicles and take public transit or engage in powering two pedals, they do more for their city than they would do for themselves. They resist the narrow, economical choice, favoring dimensions that bolster the community, which supports the urban environment, other people, the nonhuman world, and themselves.

To counter this position, one could make the argument that a person could be an ethical egoist who is only motivated by self-interest, or they could claim that an action could only count as being moral if they were doing it for some higher moral order such as fulfilling or obeying a duty. First, let us consider this point as it is in the current (anti-)framework. It is moot because, as an inherently consequentialist approach entails, we are concerned with the arrangement of parts and the outcomes that they help produce. Their motivation is almost always irrelevant. Second, speaking to the latter portion of the claim above, the view that people should have the right moral motivation to make an action qualify as moral is a "luxury" demand that flies in the face of people who are suffering. Maintaining this stance suggests that suffering should only find immediate relief if the people who are helping relieve such burdens have the right thoughts happening inside their heads. For such cases that deal with situations that range from the unfavorable to the abhorrent, having the "ideal" motivation would be preferred, of course. However, even if we could get incredibly selfish people to engage in practices that improve life for other people, the contributing impacts should not be discounted, far outweighing any such motivation.

Along with this point, it would be a mistake to ground such a position by appealing to political or social ideology, a move that would deter numerous urban inhabitants. Instead of such appeals, verifiable and measurable outcomes in public health transcend cultural, social, and political boundaries—or at least they should. Through employing these empirical measures, we gain common ground for a conversation that clearly indicates *one place* where the lines should be drawn that far outweighs what Caso could label as being closer to animal instincts.[14]

Here is one example of this outlook. Consider, for instance, that social isolationism has recently emerged as a complicated issue for public health. For instance, a newly released "systematic review of systematic reviews on the topic makes a compelling case that this subject will require advanced and interdisciplinary efforts to remedy such situations.[15] The researchers conclude the following points based on their extensive study:

> This systematic overview highlights that there is consistent evidence linking social isolation and loneliness to worse cardiovascular and mental health outcomes. The role of social isolation and loneliness in other conditions and

> their socio-economic con-sequences is less clear. More research is needed on associations with cancer, health behaviours, and the impact across the life course and wider socio-economic consequences. Policy makers and health and local government commissioners should consider social isolation and loneliness as important upstream factors impacting on morbidity and mortality due to their effects on cardiovascular and mental health.[16]

With the above passage in mind, we know that transportation, among other factors, plays a role in social isolation. Connecting affected residents with planners could mitigate these burdens. Working in concert shows respect and, perhaps, love for fellow urban dwellers and moves toward urban enlightenment. Through developing an approach to urban transportation that motivates people to use their own reasoning to think about their cities—if this were the case—urban planners could not only develop novel ways to move about the metropolitan sphere, but they could create pathways that could move people to care about their city that transcends the economics of self-interest. Such measures should push against the efforts to maintain the transportation status quo when it causes or perpetuates harm or oppression in any shape, form, or fashion, which also includes but is not limited to egregious miscalculations and justification of moral ordering for narrow self-interested thinking.

Under moral ordering, we are operating within a method that pushes against self-interest by its nature. As a technology of thought, it should be resilient to missteps. If employed as suggested, or in a manner that does not stray too far from the structure of its inner nature as argued in the previous chapters, then its operational capacity should shield against abuses and favor actions that respect moral ordering. In turn, urban planners who elect to work with residents could transition from urban paternalism to cooperation. This process is cooperative urban planning, "co-planning." If there is cooperation, then such actions can set the stage for relieving the tension because both parties stand on the same ground, developing transportation systems that move people about the city.[17] Yet, delivering such outcomes is less likely to be unjust, considering that the people that it will serve can meaningfully influence it.

This view's significance is that when travelers and transportation officials work together to create mobility systems, they are co-creating outcomes that reflect a sense of value that bring experts and non-experts together, providing congruence. This point does not suggest that the history of transportation lacks instances of these kinds of practices. On the contrary, such cases are indicative of the measures that require emulation to attain similar outcomes. Through such actions, we can square such measures with Caso's notion that we create values in concert with others as they pertain to transportation infrastructure. In turn, identifying the specific parts, or the lack of particular kinds of parts, becomes a way to approach these issues with respect in a manner that aligns with moral ordering. Bearing in mind that each city has unique characteristics that include different populations of vulnerable and marginalized people, publics, ecosystems, and urban artifacts, this reality reminds us that efforts should be customized, a notion that reinforces the utility of flexibility that is inherent in moral ordering.

Of course, sticking with the idea that there is no one-size-fits-all solution, some cooperative urban planning measures will require more work to get residents on board with professionals. For instance, if there is an extensive history of injustice, corruption, or unethical behavior in a given situation, then gaining the needed ground for conversation might require additional efforts. That is, to understand each other and to chart a way forward that deals with such realities, pursuing restorative efforts could serve mutual interests well.

For cases with different conditions surrounding them, the goal will be to create co-planning (sub) approaches that will support endeavors that align with the historically situated circumstances pertaining to such scenarios. The common theme that must hold steady here is that the parties involved are open to creating pathways, assuring that shared views are based on cooperation.[18] This idea suggests that planners and residents should engage in certain practices (e.g., restorative efforts that can mend relationships) if they want to attain this outcome.

The task for planners is to create reasonable and accessible means for people to move—and *move them to care about the means*. This claim is quite demanding, which is why it would count as a supererogatory action that calls for moral praise. For residents, once they engage in this process, they move toward urban enlightenment. While some philosophers are motivated to bring people out of the darkness and introduce them to the light of reason, the urban planner should guide people to use their reason to think about urbanity, working to remove the barriers that stymie their abilities to do so. Urban mobility, considering its encompassing impacts on people's lives and life on the planet, can serve as the first step that can lead us there. It can encourage us to contemplate moving, which is a way to philosophize about the city, which includes but is not limited to transportation issues such as bike paths, road ecology, and sustainable infrastructure. Think, move, and take it to the streets.

The above notion of urban enlightenment is only one view of what is required for building and improving the city in a manner that seeks to transform power relations in the urban sphere meaningfully. Although this order is a tall one, it is only a vehicle that can help us achieve other goals that go beyond merely getting it right when it comes to arranging and dealing with the parts of transportation systems to yield better outcomes.

Bearing this notion in mind, employing moral ordering and working to secure the conditions for urban enlightenment can support worthwhile goals such as socially just urban sustainability and urban flourishing, meaning that people can live and thrive in cities. Although the above goals are suggestive, they are reasonable ones when considering the reality that we live in an age of climate change, along with vast social and economic inequalities. This idea suggests but does not entail that the larger, mentioned goals hinge upon moral ordering and urban enlightenment, or vice versa. Instead, the point is that paying attention to such dynamics could help deliver such possibilities or other ones that align with the cities who would employ such measures to attain those ends.

The idea worth emphasizing is that producing a morally-ordered future is that people who live in cities would be able to meaningfully weigh in on the process of moral ordering through co-planning with municipal professionals. Considering that urban mobility maintains a high degree of influence, beginning there makes sense. Aside from this aspect, the "battle" for the future of transportation systems is playing behind the scenes of our cities. In turn, the future is being decided, and the questions that need to be asked and answered are these two: What kind of future do we want for urban mobility, and Who gets to decide it?

The problem with addressing such inquiries is that hardly anyone bothered to ask the questions before moving forward with the steps that would serve the implementation of a future that has the primary interest of private companies in view. Meanwhile, transportation experts are shouting from the sidelines, advocating for a future for the betterment of our cities. One could argue that they do not have a good public-relations department when compared to industry.

By sticking with the theme of urban enlightenment through moral ordering, people thinking for themselves, and thinking in concert with municipal transportation officials, the following chapters work to advance this view. They examine the possibilities of having worthwhile goals and many of the anticipated futures of urban mobility, along with alternative futures that serve the ends of many worthwhile goals, as mentioned above. While such an exploration endeavors to make a case that holds steady in its partiality for urban travelers, it does maintain a position that is inherently antagonistic to enterprises that aim to *only* make a buck. Instead, it seeks to foster the conditions that allow such outcomes to manifest when they align with the communities they will serve—rather than using residents to increase their profit margins.

NOTES

1. David Harvey, "The right to the city," *New Left Review* 53 (2008): 23.
2. This idea also needs to consider highly significant aspects as systemic racism, classism, and sexism that need to be removed from the institutional and social structures of cities, states, and nations.

3. There is recent paper that develops and advances how we weigh time in urban transportation, underscoring the significance of this line of thinking. For more information, see Maria Nordström *et al.*, "Let me save you some time . . . on valuing travelers' time in urban transportation," *Essays in Philosophy* 20, no. 2 (2019): 206.

4. OPALportland, *Adela's journey*, 2012. https://www.youtube.com/watch?v=IjxebYmey44.

5. OPAL Environmental Justice Oregon, *OPAL environmental justice Oregon*. https://www.opalpdx.org. To make a donation, please visit https://www.opalpdx.org/make-a-donation/.

6. There are also issue that fall outside of the immediate areas of concern that also require examination to see how they remain relevant, even if such relevance is tangential at best. For instance, damaged roadways could also impact public safety.

7. Commenting on the details of the technical operations here is not my area of training and is best suited for the training that professionals receive the necessary education to deal with these practical affairs.

8. George Russell, Irish Agricultural Organisation Society, "Notes of the week: Fair play in legislation," *The Irish Homestead: The Organ of Irish Agricultural and Industrial Development* 17 no. 53 (1910): 1087.

9. Michela Le Pira *et al.*, "Competence, interest and power in participatory transport planning: Framing stakeholders in the 'participation cube'," *Transportation Research Procedia* 48 (2020): 2386.

10. John LaPlante *et al.*, "Complete streets: We can get there from here," *ITE Journal* 78, no. 5 (2008): 24.

11. Immanuel Kant, "Answering the question: What is enlightenment?" *Berlin Monthly. Berlin: Berlin Monthly*, 1784.

12. This point means that I am departing from Kant's original context (i.e., "freestyling").

13. Kant, "Answering."

14. I do not mean to insult animal-enthusiast for my portrayal of animals as lower-order beings, considering that many nonhuman species would never engage in many of the destructive, mindless practices that humans commonly engage in and defend in a strongly anthropocentric manner.

15. Nicholas Leigh-Hunt *et al.*, "An overview of systematic reviews on the public health consequences of social isolation and loneliness," *Public Health* 152 (2017): 157.

16. Leigh-Hunt *et al.*, "Overview," 158.

17. One could say that co-planning is participatory planning, and I do not have any significant problem with that claim. However, participatory planning seems to suggest that people are *allowed* to participate. Co-planning embodies an attitude of mutual respect wherein residents are seen as more than people who have permission to participate. Besides this point, "co-planning," one could argue, has an aesthetically pleasing sound, which could motivate people to work with transportation professionals. So, there is that.

18. It is worth mentioning that cooperative urban planning is not simply another way to speak of participatory planning. There are at least two central distinctions.

First, cooperative planning goes beyond the notion of merely participating, but it shows that mutual respect is at play, even if such respect is minimal. That is, all cooperative planning involves participation, but not all participation involves cooperation. Two people can participate in an activity without respecting each other. However, cooperation (in this sense) entails mutual respect to advance a shared goal. Each party must acknowledge respect for the other in a way that can lead to the kind of attitude that can deliver a flourishing way forward for the city, residents, and professionals. Second, there can be specific kinds of cooperative urban planning, one that is indicated in the chapter is restorative. The ability to create this kind of effort illustrates that it is possible to create approaches that serve a particular situation. Future research endeavors should focus on developing specific ways to address complex planning issues in a given city.

Chapter 9

Moral Ordering and Worthwhile Goals

Here is a summation of the previous chapters that connects them to the message of this one: there are transportation systems that, through the part's configuration, have become harmful. *Moral ordering* is a process that can help alleviate harm through creating pathways to cooperative urban planning that benefit the stakeholders as described in the moral ordering process. In turn, people could begin to think for themselves on these issues, which is a step on the pathway to urban enlightenment. It is a journey that can help metropolitan environments transform into co-planned cities that are not guaranteed to resemble the past or the present. These conditions open up the possibility for a better future. Cities that adhere to this description are places where humans can work toward worthwhile goals such as socially just urban sustainability and/or human flourishing. The point here is not to limit the kind of worthwhile goals, only to show that efforts such as co-planning, guided by moral ordering, can support more substantial aims that could improve life on the planet.

Now that we have had a quick recall of the totality of claims on urban mobility and their trajectory, here is the current situation that makes urban enlightenment a pressing yet complicated affair that will require attention in several ways. The reality is that there is a lengthy history of leaving communities out of decisions in transportation.[1] Even though some municipalities and transport authorities have made impressive strides toward inclusive planning, it remains a challenge in many places today, meaning that marginalized groups still suffer.[2] This claim indicates that if we want to radically change these conditions, addressing two primary elements can create possibilities that can deliver better outcomes.

Recalling from a few previous ideas, the first is to create a way to address the moral considerations for the sometimes conflicting groups of stakeholders, which include vulnerable and marginalized populations, the public, nonhumans species, future generations, and urban artifacts, through what has been called *moral ordering*. It is a process that guides the prioritization of the decisions that will lead to actions that will benefit or harm these categories, beginning with serving the people who are most impacted by such decisions, due to a variety of considerations.

This notion requires that any decision for the public must not harm but benefit people in the first category. When it comes to weighing competing intra-group interests for the public, we should employ the notion of respect to avoid sacrificing the group for the individual while at the same time not putting so much stock in the interests of the individual that it harms the group.[3] Considering that we have strong interests in building transportation systems that are inherently anthropocentric but that should not disregard the intrinsic and instrumental values of nonhumans, we should aim to deliver solutions that bring these matters to the forefront of our thinking.

We want to keep future generations in view, but they do not exist yet, so it is challenging to say that we owe debts to them. However, we have an interest in their existence, meaning that we owe it to ourselves to preserve the conditions for them to lead genuine lives. This notion suggests that we should ensure that our decisions to enhance urban mobility should not harm them in a significant capacity. Lastly, because urban artifacts such as buildings, bridges, landmarks, and neighborhoods matter but are not as valuable as human life, they fit at the end by default.

While this process seems absolute, it is only suggestive, allowing for flexibility when there are strong reasons to waver from the order. Additionally, this method must include respected views from the stakeholders who are going to be affected. This point leads us to what happens when people start to examine and debate these aspects. Such a process begins the journey that could lead to urban enlightenment: people thinking for themselves on mobility affairs in the manner described above. Accepting this notion with any degree of seriousness requires that we decrease the need to have strongly paternalistic and hierarchical transportation planning. Specifically, we can call it urban paternalism. One motivation that underpins such concerns is that we cannot usher in a new process while all the parts are moving. So, transportation planners need to develop meaningful measures of inclusivity that respect people and themselves (as trained professionals). In turn, they can establish reciprocity that can create pathways to urban enlightenment for both parties.

Here we return to the notions of respect and balance. It must be done in a way that keeps the trains running on time. Yet, it includes the possibility that people can weigh in on such procedures when it does not result in catastrophe. We do not want people to harm themselves due to a lack of specialized knowledge that prevents injuries and calamities. This notion opens up the possibility that people could want what would otherwise be called paternalism, once they see the risks that remain inherent to transportation planning and engineering. If they decided to give up their voices once they understood the dangers, they would be abnegating to paid professionals who can free up space for them to pursue flourishing through other ways. Fulfilling such conditions provides a way for people to participate in shaping urbanity. The choice is theirs or at least that could be a possibility. Still, people could see their stakeholder-ship and want to claim it. In turn, this process could create opportunities for these kinds of endeavors, and transportation systems, as argued, come with significant reasons for urban dwellers to get involved.

If people can transform areas such as mobility systems, then they can take back power, if only minimally and incrementally, remaining compatible with the insights that the Lefebvre-ian traditions have grounded. Although such actions show how citizens can take a piecemeal approach to reclaim specific dimensions of urban power that speak to different personalities, cities require managers and departments to maintain a city's daily operations. How, then, do we allow residents to become meaningfully engaged citizens—if the municipal apparatus has inherent limits due to its arrangement and bureaucratic and technical nature? This notion suggests that there must be operations that necessitate officials who work on behalf of the city, and—at the same time—there is no reason why outsiders should not be included when it is safe, reasonable, and feasible. Embracing such an approach can help establish the kind of cooperation that is needed for urban enlightenment. In terms of the required cooperation, each party has specific tasks that they must undertake, which does speak to the far-reaching need for consistency for cooperative planning. From the professional side, attending to stakeholders' needs over the stockholders' motivations should hold steady.

For instance, we do not want special-interest groups to persuade, bribe, or con residents into wanting something that goes against their ability to maintain safe (or just, fair, moral, etc.) living conditions that are clouded by short-term interests. Viewing such positions through a lens of specialization with an inherent sense of paternalism aligns with the traditional attitude toward overseeing municipal operations. Such a concept entails that some areas are off-limits to community members, such as weighing in on building codes based on elements such as science or matters of public health.

Yet, without questioning if this is an explicit course of action wherein the conditions of urban management demand exclusive control by specialists, pursuing only such a course could amount to dogmatic thinking. Bearing in mind that creating cities' identities will take concerted efforts, municipal officials who make a living from city residents must aim for, or at least meaningfully and fully explore, inclusivity in such undertakings. At the least, they must not impede residents' pathways to urban enlightenment.

Although this attitude could maintain the kind of cities that exist today, what if we want different types of cities—ones designed for socially just sustainability? In turn, we could establish new standards of urban living, ending the pattern of creating places to live that harm people in myriad ways. Suppose we can break this cycle, one city at a time. We could contribute to the (global) environmental science that is needed to avoid the kind of environmental calamity that concerned and motivated Jonas, and we could go further to create cities that strive for worthwhile goals.

Bearing this notion in mind, we can create different kinds of cities, a reality that stems from creating the possibility that people can *actually* have a voice that strays from the status quo. Giving this thought the attention that it deserves requires that we make radical changes that can create cities so vastly different from today's models that we might have to come up with new names for these habitations of human enterprise. This point suggests that although we need planners, engineers, and architects who can deliver the technological solutions that can alleviate harm, professionals who can cooperate with residents to create the kind of cities will shape residents in the ways that they want to be shaped. When people participate in their cities in such manners, they co-plan places wherein they determine the conditions for the worthwhile goals they set.

If these ideas lead to real-world change, then the bold professionals who undertook such challenges deserve moral praise in select instances.[4] The urban residents who served as informal leaders who would accompany such efforts would become vital players who deserve the highest respect. The following section below examines them in detail to flesh out these points, moving toward an expansive conception for such conditions that could apply across a wide range of similar situations.

TRANSPORTATION SPECIALISTS, MORAL PRAISE, AND RESHAPING URBAN MOBILITY

The ideas explored in the previous section show that adopting novel measures such as co-planning initiatives will require specialists and residents to venture into territory that faces several kinds of impediments together. From naysayers who defend the same norms shown to be harmful to the practical realities of making mistakes that come with instituting new practices, concerns are not lacking going forward. Further, the stakes are high for stakeholders. They bring every area of urban living into question, risking the sense of security that we commonly associate with their familiarity.

For example, when tinkering with the ideas that pertain to transportation in the city, we cannot dismiss the notion that mobility connects everyone with the means to survive. Consider workspaces, commercial districts, and neighborhoods. Due to these parts' arrangements, mostly being spread apart from various distances, the need to travel goes without saying. Buildings are central to urban living. We live apart, but then we need to be together as our lives require. Then, we must return to our buildings. With these points in mind, it is evident that movement is not done in isolation when concerned with urban mobility. Such a notion entails that we are doing it in the city. Moving for its own sake is one thing, but we are talking about moving with purpose. This idea does not entail that we cannot travel through the city with no particular destination in mind, but most local trips that people take are to complete a task. In turn, urban mobility is a means in most instances, and the means should not deter us from seeking the ends.

However, as shown in the previous chapters, the ends are places of employment in most cases, meaning that, for working folk, they must endure the trek or forfeit everything else. Speaking rhetorically, building in the city requires a means for movement. Moving in a manner that does not create harmful conditions, as illustrated previously, requires intensive thinking. This notion is not meant to suggest that people who made transportation systems were not thinking. However, when considering the outcomes that the experts did facilitate, those realities result from their thinking.

Robert Moses was thinking. Transportation scholars have argued that his vision has shaped numerous cityscapes, meaning that he played a dominant role in the outcomes that burden urban dwellers and transportation specialists alike.[5] When I say that Moses took on such a role, that phrase has two meanings. First, we are dealing with Robert Moses, the person, and we are also concerned with the transportation specialist who happens to be named Robert Moses. The former is no doubt the subject of moral praise and blame for his personal actions. In terms of the latter, however, that notion requires a bit more fleshing out.

That is, titles representing professions such as "transportation planner" and "transportation engineer" are inventions. They are things that were invented to handle particular tasks. If we think about them in this manner, then these specific titles are kinds of technologies—conceptual devices that we employ to get the job done. Although they drastically differ from the sort of technologies that we typically encounter, use, and, in an ethical context, analyze, people still use these technologies to accomplish tasks that have real-world effects. Without the job title, Robert Moses, the person, cannot build highways through cities. He cannot force people to move so that crews can construct highways. It could be the case that construction workers build stretches of highways and bridges without knowing the name of the person who drafted the plans, but they knew that someone with a relevant job title did make a plan. Robert Moses happened to be that person—the one who used his twelve official job titles to create layouts of places like New York City.[6]

If we think about occupations as technologies, then it becomes clear that we can show that they played a moral role in larger socio-material arrangements. This point entails that, like physical technologies such as vehicles, infrastructure, or laws and policies, the job titles that we find associated with such projects also play a moral role in good or bad outcomes. To be exact, they are passive parts directed by an active part. In mereologically inspired terms, they count as abstract parts. While these different technologies fit broadly into the same category of "technology," we also need to show that they remain dissimilar. In such cases, we can say that material devices are passive technologies because they occupy passive roles in socio-material arrangements. However, for job titles, to show that an individual only "uses" them, we can make the additional distinction that they play passive moral roles in those arrangements. They qualify as passive, abstract parts of a transportation system. In turn, we can say that a transportation planner played a moral role in creating such results.

By looking at these outcomes through this lens, a view emerges wherein we can address the harmful or beneficial outcomes that such specialists play without engaging in building a case that they were immoral or moral persons. Making such claims requires a professional who was assigned such a title either misused, willfully engaged in inherent wrongdoing, or engaged in supererogatory actions. With this point in mind, questions emerge that aim to illustrate that these points require criteria. For instance, what kinds of conditions are needed to claim that transportation specialists who are merely doing their jobs require moral praise?

While numerous cases could require moral praise when examining specific scenarios anecdotally, there are a few broad situations wherein such instances require attention. For example, when addressing and making progress toward multi-tiered issues such as those that result from climate change, systemic discrimination, and impossible interwoven challenges, crafting mitigatory efforts requires a lot more work than simply making the trains run on time. From an outsider's perspective, one can only imagine the amount of brainpower that goes into dealing with such complexity. Yet, when they discover a novel solution to a particular problem that brings all of the above elements into perspective, their work is not seen as moral. Although they often receive praise for successful efforts in placemaking, such accomplishments need to produce moral outcomes. By acting as active parts, these professionals directed the passive parts, which include parts such as job titles, streets, buses, and bike lanes, producing repeated, predictable situations that will improve people's lives. Bringing people into the behind-the-scenes process provides a way for residents to help shape the city and their lives.

One immediate worry is that every time that a transport professional gets it right, she will require moral praise. Due to the extreme frequency of use, giving moral praise will become meaningless.[7] Avoiding this situation requires gesturing toward standards that make giving praise genuine or establishing criteria for when it is fitting.[8] Considering the "anti-framework" framework theme that runs through this book, sticking with that pattern provides consistency. Still, bearing in mind the intricate nature of moral ordering, paying attention to inclusivity to produce good outcomes in urban mobility is a reasonable way to look for cases that deserve moral praise. This notion does not entail that transport professionals must co-plan solutions for wicked problems that satisfy every group in the moral ordering process. However, that kind of outcome would strongly warrant investigation to see if it does deserve moral praise.

The point above also does not strictly hold that all groups must receive moral praise. Still, considering that some incredibly demanding mobility affairs could require dedicated, advanced efforts that go far beyond the customary expectations for people who engaged in those positions, studies that focus on such matters require examination. Further, for an example of on-the-job moral leadership, bus operators who occasionally go beyond the call of duty for protecting victims of harassment, disregarding protocols (when safe) to help disabled persons such as stopping at undesignated places to provide practical assistance, and serving as a social pillar of the mobile community deserve accolades.[9] These people earn moral praise for providing the necessary but not required theater for numerous actions that go unnoticed.

These kinds of instances would need examination on a case-by-case basis. Despite lacking a clear-cut measure to determine when moral praise is appropriate, employing the sentiments showing that there are cases that would be enhanced by moral praise also reveal societal values. By expressing them in the form of moral praise, societies gain a manner wherein they can exhibit a complex understanding of the advantages/disadvantages associated with transportation systems. This notion is of paramount importance when considering that we are dealing with people's ability to lead flourishing lives.

URBAN RESIDENTS, MORAL PRAISE, MORAL RESPECT, AND RESHAPING URBAN MOBILITY

When it comes to the conditions surrounding urban residents and their roles in transportation affairs, they will drastically differ from professionals, along with the requirements for moral praise. Despite these differences, such realities do not dismiss the possibility that residents can engage in everyday actions that warrant moral evaluation to see if they qualify for moral praise in a way that bears a resemblance to the pattern that underpins moral praise for professionals. These instances account for the other side of cooperative urban planning measures. If specialists deserve praise for the work inside city halls, then residents might also require it for the efforts on city streets.

For instance, engaging in co-planning will require a significant commitment on their parts. Due to such actions' unconventional nature, they must be brave to undertake tasks that could fail or be a waste of their time. In turn, it is fathomable to hold that many people will not have the availability to do so. Still, there will be individuals who make sacrifices for the betterment of urban mobility, and the character of such acts sets them apart from the ordinary, suggesting that they might deserve additional study to see if they reveal insights into how we think about transportation and morality. While non-experts lack the technical training that can prevent tragedy, it seems impractical to believe that they should be held to the same standards of moral praise as transportation professionals. Yet, it seems reasonable that some cases hold steady as strong contenders for moral praise. For other such instances, a lesser form of praise could be in order.

First, people who have fought for transportation justice deserve moral praise, a point that should be axiomatic. Individuals who made their lives revolve around transportation for reasons that transcend functionality could also deserve moral praise for efforts that have measurable outcomes yet will never be counted. For the people who have or who would work with transportation professionals to improve mobility services, such cases should also be examined to see if moral praise is applicable.

When it comes to residents and their daily involvement with transportation systems, we have to keep in mind Caso's point about doing more for others than you would have them do for yourself. While such actions are not always supererogatory as we saw in the case with urban planners, such actions belong on the spectrum of moral outcomes. While it might not be necessary to say they deserve moral praise because it is not the same thing as above, we should call this moral respect because their actions require accolades that transcend regular respect. We could say that they deserve moral respect. They deserve a higher degree of respect than the way we commonly employ the term, and moral respect serves to acknowledge the moral dimensions of their roles as active parts and can affect passive parts that help create or support moral outcomes.

Such measures could include an array of actions that remain germane for the modes of mobility that are available for travel, meaning that associated actions or nonactions are subject to the circumstances surrounding them. For example, some conditions for moral respect could include picking up trash from the floor of a bus or giving up one's seat to a passenger who just seems tired. On roadways, drivers who *always* use their turn signal or kindly let other motorists merge could warrant moral respect. These sorts of actions, while they should be customary, are not, and they vary from one place to the next. Giving moral praise for them seems to go too far, rending the practice meaningless. Moral respect, however, serves as a way to balance such practices.

These points could prompt one to ask: What does giving moral respect to someone look like when engaging in such a practice? The answer is basically what does happen when someone engages in those actions. For instance, giving up one's seat on a packed subway car or bus could solicit an approving nod of the head from a neighboring passenger. It could be receiving a smile. Introspectively, it could be the inward approval that one gives one's self. The point here is not so much to focus on how moral respect is observed, but it has more to do with the conditions surrounding it to provide ways to encourage it. By developing a way to employ "moral respect" as a conceptual device for attending to such matters, we could develop real-world measures that encourage behaviors that would garner moral respect. In a certain sense, many of these measures already exist in a negative form, serving more as a deterrent to unwanted action.

For instance, at present, there are signs on subways and buses that encourage particular actions such as "remember to reserve these seats [towards the front of the vehicle] for passengers who need them" and "no cell phone use allowed." These signs imply that failure to adhere to appropriate behaviors is frowned upon, which borders on rudeness. Although committing such actions is not inherently immoral, they do gesture toward unpleasant outcomes. One could easily imagine a case wherein a young man could be shamed for not giving up his seat to a woman in her ninth month of pregnancy. On the opposite side of the scale of possible actions, we could also imagine a sign that encouraged actions that would solicit moral praise. Such instances could remind people that giving up one's seat for a tired passenger is a good thing or that checking for one's left-behind rubbish is a desired action. These kinds of gestures could, continuously in concert with each other, help deliver better outcomes for passengers.

One could make a case that transportation specialists could also deserve moral respect for behaviors that can lead to similar outcomes. This objection is fair. While they deserve respect in a general sense, their employment introduces an element that is missing in the case of transportation users. For instance, they are expected to engage in the task of transport planning because they receive compensation. They would be relieved of their duties if they failed to do so. Transit riders and commuters are only expected to follow laws and rules. Any other actions are voluntary. Due to this difference, making a case that transportation professionals deserve moral respect must account for this discrepancy. Due to this condition, it makes sense that we stick with moral praise for them in select instances while extending moral praise and respect for transportation users and drivers in the cases that warrant it. However, when they are off the clock, riding as bus passengers, then we could reintroduce moral respect.

With these novel elements in mind, we see that there are motivating factors that can move us to improve urban mobility, the moral means to do so, and ways to usher in such actions to create a future for transportation systems that can bolster worthwhile goals such as socially just urban sustainability for human flourishing. Yet, this work is merely a sketch that should point us in the vicinity of the efforts that are needed to move from the world of ideas to the streets. The previous pages touch on several areas, philosophical and practical, that will require advanced development. In the closing section, I explore such possible avenues that can help us get the job started.

FUTURE AREAS OF RESEARCH: FROM THE ARMCHAIR TO THE STREET

Although the trajectory above seems straightforward, analyzing transportation issues by using inclusive moral ordering to deliver better outcomes suggests that there is a chance that it will carry over in the real world. This idea underscores a common sentiment: "It works in theory, but what about in practice?" If one were to argue that the latter is the true test of the former, then crafting optimal measures for executing such an order will depend on efforts that must originate from outside philosophy. Several neighboring disciplines whose scholarship could help propel some of the ideas were presented in the preceding chapters.

Perhaps the area that requires the most development is creating a way to enhance the municipality's integrity so that it can foster inclusivity in sophisticated spheres such as urban mobility. Moving forward with the changes that can help us create better streets and cities cannot rely on past practices that have failed to deliver such outcomes. We need measures that are truly radical in the sense that they seek to supplant previously employed methods. The concern here is that, when seeking to find a replacement that can get the job done, there are few visible contenders. This idea suggests that we are dealing with an essentially radical affair, and it needs attention. The penalty of ignoring this reality is that other people who stand to profit handsomely from urban mobility would largely determine its course.

At present, such parties include industries that deal with automated vehicles. This notion is not surprising when considering that reshaping the sphere of urban mobility not only requires a substantial investment from cities, states, and nations, but it also directs unimaginable and unending sums of money from most urban travelers the world over. Due to this enormous shift, which could arrive in a piecemeal fashion, there is a great need to examine the many moral dimensions that remain connected to replacing numerous parts in a way that could be permanent. In turn, chapter 10 heads in that direction, examining several of the facets that pertain to such conversations and our understanding of such a possible reality.

NOTES

1. Richard Ezike *et al.*, "Defining "communities of concern," *Transportation Planning: A Review of How Planners Identify Underserved Communities* (2020): 9.
2. Richard Ezike *et al.*, "Defining," 12.
3. By saying "the public," I do not mean to say that vulnerable or marginalized people are not members of the public.
4. I first explored the idea of moral praise for urban infrastructures elsewhere, wherein I go into much greater detail. Specifically, I explore the idea of professionals such as the former mayor of Atlanta, Shirley Franklin and the conditions for her and people engaged in similar professions to receive moral praise. For more information, see Shane Epting, "Urban infrastructure and the problem of moral praise," *Techné: Research in Philosophy and Technology* 25, no. 2 (2021): forthcoming.
5. For example, see Robert Caro, *The power broker: Robert Moses and the fall of New York* (New York: Alfred A Knopf Incorporated, 1974).
6. Sydney Sarachan, "The legacy of Robert Moses," *Need to know on PBS*, January 17, 2013.
7. Shane Epting, "Urban infrastructure and the problem of moral praise," *Techné: Research in Philosophy and Technology* 25, no. 2 (2021): forthcoming.
8. Ibid.
9. For an example of such instances, see Shane Epting, "Ethical requirements for transport systems with automated buses," *Technology in Society* 64 (2021): 101508.

Chapter 10

Thinking, Moving, and the Future

The case for moral ordering provides strong reasons to believe that thinking about the arrangement of transportation parts could be part of a larger, more moral, whole transportation system. It is one that urban travelers could help plan, giving them a sense of ownership in some cases, or perhaps they could see a bit of themselves reflected in the structure in other instances. Then, the future of urban mobility holds the possibility that it might not resemble the past closely, meaning that it could address existing problems while paying attention to anticipated needs. If we hold such notions as realistic undertakings in some capacity, then worthwhile goals could become attainable dreams, perhaps sooner than we could imagine.

These points aside, when we look at the predictions for urban mobility, the loudest voices are the ones shouting for a future of automated vehicles (AVs). However, considering the enthusiasm behind such calls, it says that a future for AVs' sake could be a more accurate description. In brief, the gospel is that, due to their highly sophisticated nature, there may not be a need to consider having to turn to advanced thinking because technology will eliminate the need to make such complex decisions. It sounds incredibly exciting, calling for our imaginations to envision a "world of tomorrow." With this point in mind, let us entertain a futuristic scenario.

In this dreamland, there are not any transportation inequalities. Everyone is ushered through the city, never having to stop from their homes to their destinations because the vehicles communicate through 17G speeds. They can slow down or accelerate accordingly. Seatbelts? There is no need for them when the vehicles never crash, and there is no need for parking because they cruise around on magic motors that we power through our smiles. Terms such as "driver's seat" become obsolete. Everyone can sleep, study, watch films, communicate with people far away, or simply enjoy the neon city as they silently pass by people in pleather outfits. While this scene sounds completely outlandish, it is not that far removed from the realities that several academics and industry leaders promote regarding the future for AVs and urban mobility.

That is to say, when examining claims about the future and AVs, an "assumed view" emerges, holding that this technology has the possibility of solving numerous issues in transportation. In turn, we are left with somewhat of an utopian view of urban mobility. While I cannot speak for Jonas, it seems that the promise that these vehicles hope to deliver would support his imperative (ironically). This point suggests that it might be the right decision for municipal leaders to make transportation plans center on AVs. However, the manner wherein they are discussed qualifies them as parts intended to serve a utopian vision of hands-free and mind-free mobility. Yet, the question that we need to ask is this: Whose utopia is in the works?

It is assumed that people want such a future. Still, there are significant reasons we should question the tacit motivations underpinning the hoopla, but posing such inquires could give the impression that anti-technological bias is afoot. That is not the case. Instead, the central idea is that the city's immediate or near future rests on the premise that it should belong to the people who are the essential parts that give life to the urban sphere. This notion suggests that we should examine the transportation parts that will largely shape the bigger transportation whole through moral ordering, which will, in turn establish the contours of people's lives.

While there is little doubt that we should champion advancements in research that can alleviate transport-related harm and improve urban mobility, there are several reasons for people to remain critical, employing moral ordering to learn how each stakeholder group could likely be impacted. For instance, the uncertainties that come with the possible impacts, the time scale of implementation, and unknown challenges that pertain to AVs give us reason to have reservations about making all other transportation decisions revolve around driverless vehicles.[1]

Although we do have good reason to believe that AVs will eventually launch, we ultimately cannot predict their impacts or when such effects will manifest. Instead of embracing driverless vehicles with trusting gusto, transportation professionals should consider them as another component of transportation planning—instead of *the* transportation plan. In addition, the manner wherein they are introduced into the cityscape should also hold steady as an interactive affair if people are to gain a sense of ownership that could accompany co-planning. One of the significant challenges to understanding the issues that will emerge in this domain is that there is no precedent to cite.

Despite this shortcoming, recent instances of technological implementation can serve us in such an endeavor, namely transportation network companies.

Investigating how these parts landed in the larger transportation whole provides us with clues about what we could expect to happen with AVs and their role in creating someone's mobility utopia. This idea is not intended to suggest that we should rally against AVs. Instead, it wants to identify ways that, through moral ordering, can create the possibility to introduce these driverless parts into a larger whole in a manner that facilitates urban enlightenment so that it can support worthwhile goals, as mentioned at the end of chapter 9. It is one wherein urban residents can use their reasoning to deliver a future they want in concert with honest planners.

To make this case, I will examine some of the predictions made about transportation network companies (TNCs) that offer private mobility services. Then, I exhibit several of the claims that have been made about AVs, arguing that both of these technologies rest on forward-looking arguments. Due to such positions' likeness, one can make a strong case for following the advice of transport planners who remain focused on the best way to move people about cities without dogmatic adherence to technocratic idolization. In closing, I explore some notions that could benefit cities in delivering improved mobility services to urban residents.

THE ASSUMED VIEW OF AUTOMATED VEHICLES (AVS)

To get an idea of the full range of the anticipated effects that driverless vehicles could have on cities, several researchers have argued that their impacts will completely reshape society.[2] Due to these predictions, numerous papers that discuss AVs are engaging with how these technologies could improve transportation, along with several issues that deserve attention.[3] Such challenges are present in many of the explorations of driverless vehicles, which are not unique to any specific research area, including fields such as engineering, urban planning, and philosophy. Indeed, AVs are an area of investigation that could benefit from an interdisciplinary pooling of resources for which Jonas advocates.[4]

When it comes to the ethical aspects associated with programming these vehicles, philosophers have thoroughly explored a myriad of issues, indicating numerous problematic areas.[5] Yet, one could argue that these philosophers address these ethical issues because these technologies will emerge, and they will become ubiquitous due to AV's many perceived advantages.[6]

While such works advance our understanding of the moral dimensions that pertain to driverless vehicles, the point that I want to make does not involve debating any specific position that has been established in the literature, ethical or otherwise. I am not targeting any particular argument in what follows.

Instead, I exhibit a general trend in talking about AVs, which consistently discusses their possible benefits across a wide range of disciplines, despite possible challenges. This point is obvious when industry leaders praise the assumed future of AVs.[7] The lure of these alleged advantages shows why the industry is trying to create this technology. The goal behind examining the assumed view is to reveal a problematic pattern in the grand sense of how we typically think about driverless vehicles' beneficial dimensions that *might* manifest in the future.

Although it is rare to find research articles or manuscripts that are entirely unwavering in their support of AVs, when examining several such works collectively, this assumed view emerges.[8] It shows that these technologies have numerous advantages for urban centers, as described below, even despite obstacles. In turn, it would seem foolish not to pursue them with great enthusiasm. Most of these benefits appear as solutions to social and environmental problems that have been mainstays within typical transportation systems. Bearing in mind that these topics concern human health, quality of life, the public domain, and environmental issues such as climate change, it makes sense to look toward new and emerging technologies to solve these problems. AVs show great promise in this regard.

Consider, for example, many scholars hold that AVs have great potential for safety, improving air quality, and decreasing crashes due to errors in drivers' judgment.[9] They also illustrate how AVs could lessen the demand for urban land use and provide a way for vulnerable populations to access needed social services.[10] Several researchers argue that driverless vehicles could help build sustainable infrastructure and communities, support urban development, and reduce overall energy consumption.[11] Philosophers who have studied AVs' predicted social dimensions argue that numerous researchers focus on topics such as how driverless vehicles could benefit public health and easing congestion on roadways.[12] Some researchers figure that AVs will significantly reduce greenhouse gases, estimating a 40–60 percent reduction in some instances.[13] In other cases, experts explore numerous possibilities of driverless vehicles, from economic benefits to energy security to helping populations with limited incomes achieve mobility. We can also focus on particular groups of vulnerable people. For example, some argue that AVs can help senior citizens increase their mobility.[14]

The above references are by no means an exhaustive list of the benefits that driverless vehicles could provide, but they represent the kind of thinking that one could argue is present when examining the possible (positive) outcomes that could happen by introducing AVs into population centers. Hence, this reason is why I refer to the above collection of claims as the assumed view of AVs. Although it is not wise to put all writings on AVs under the same umbrella, the implied notion is that we should adopt these technologies because of the results they might produce, often in spite of any accompanying ethical challenges and/or downsides they might bring with them.[15]

Yet, in estimations regarding their abilities to provide the above outcomes, driverless vehicles remain subject to the same criticisms that we find with consequentialist approaches in general. Briefly put, consequentialist positions in ethics maintain that the consequences of an action determine if it is considered right or wrong, and a common objection to this view is that it is impossible to know the future.[16] Specifically, the problem with appealing to AVs' promise as solutions to some or all of a city's mobility problems is that there is no guarantee they will have any of these effects, either partially or fully. Moreover, they could make things dramatically worse due to the many uncertainties that municipalities will face when implementing AVs onto the streets. An essential hurdle to knowing if these technologies will provide the predicted outcomes is that there is not a way to tell how they will fit in with existing transportation systems and other elements such as housing they indirectly impact. Strict adherence to the assumed view of AVs takes attention away from worthwhile goals such as human flourishing and/or urban sustainability.

While these two reasons suggest that implementing AVs into transportation systems should be carefully executed, they do not get at a fundamental assumption of modern technology that Jonas previously identified. Namely, we must rid ourselves of the utopian idea that technology without limitation is inherently good, holding that we should balance such an outlook with a fear of an ecologically unsound dystopia that could imperil humankind.[17] Yet, the fact remains: we must rely on new and emerging technologies to safeguard us from the harms of our old transportation systems. However, we cannot depend on clairvoyance when it comes to driverless vehicles—even when scientific expertise and/or predictive modeling supports it. We cannot foresee how they will fit in with existing cities and transportation systems. It would be naïve to think that introducing a new part into an existing system will not disrupt it, along with the idea that such an implementation remains entirely predictable.

One can argue that the best that we can do is to examine precedents, searching for cases in the past that might inform the future. The unfortunate reality, however, is that such a precedent does not exist. There are no other examples of driverless vehicles in history. Yet, we can examine similar events to discover how they fared, such as TNCs that deliver private, personalized mobility services. At least in these cases, the user-as-driver was replaced as a kind of part, even though it was with another driver-as-operator as another kind of part. Due to this condition, I will examine the outcomes that we saw with TNCs to serve as a close-but-imperfect precedent for AVs in the following section. Although these cases differ, a comparison should provide insights into the nature of the kind of problems that transportation professionals must deal with when wrestling with a future that includes driverless vehicles.

In turn, one could argue that these experts need to share this reality when co-planning with residents because it provides a view into the outcomes that could result from placing too much stock in a decision that remains unproven but companies that stand to profit from AVs' wide-scale implementation pursue them with great enthusiasm. This idea does not exclusively hold that making a buck should qualify a motivation as slated for exclusion. However, it should be noted that there is a contending view holding that the people's best interest or the ability to pursue worthwhile goals will have to compete with other factors. This aspect is worthy of consideration, and co-planning efforts should weigh them in the same manner as other elements when the need arises.

TRANSPORTATION NETWORK COMPANIES (TNCS) AS IMPERFECT PRECEDENT CASES

While traditional, private mobility options such as taxicabs could be considered too costly for everyday travel, TNCs are relatively cheaper, providing an alternative form of transportation from buses, subways, and personally owned automobiles. They were not completely new parts of transportation systems in a way that challenged people conceptually. However, the app that would make them accessible and the arrangement of a private driver and private automobile qualified them as new kinds of parts of transportation systems.

This point aside, early support for TNCs that offered ride-sourcing services was predicted to have several benefits, exhibiting that there were strong motivations for their acceptance as part of the cityscape. For example, writing one of the first papers that examined TNCs held that "supporters view ridesourcing as part of a suite of transport options that serves previously unmet demand for fast, flexible, and convenient mobility in urban areas. By providing an attractive alternative to driving, these services can potentially reduce auto use, ownership, and environmental problems."[18]

This passage shows that early predictions indicated that advocates for ride-sourcing companies were already endorsing an approach with a consequentialist bent—a forward-looking argument. That is to say, through focusing on several positive outcomes, there is an implied argument that, with some unpacking, could easily support policy decisions that favor TNCs. Yet, legitimizing such claims depends on delivering outcomes that matched the initial declarations for their support, which failed in some ways and succeeded in other ways. A noteworthy aspect is that these new parts eventually became integrated parts of transportation systems as whole parts. In turn, companies such as Uber became such mainstays that they began offering subscription services for users who largely depended on them for urban mobility.[19] This notion shows how these app-based transit services were highly successful in some regards. Aside from this point, if TNCs would have lessened the number of vehicles on the road, reduced harm to the environment, improved air quality, and decreased the cost of urban living, then predictions about their positive effects would have held.[20]

Despite holding such promise for improving urban mobility in several regards, researchers argue that TNCs have harmful consequences on cities and urban dwellers.[21] For instance, due to the sheer number of ride-sourcing vehicles on roads in San Francisco, TNCs did not positively affect transportation conditions.[22] A recent study holds that ridesharing services have increased the number of severe traffic accidents and fatalities.[23] In a report by Bruce Schaller, he argues: "Shared ride services such as UberPOOL, Uber Express POOL and Lyft Shared Rides, while touted as reducing traffic, in fact add mileage to city streets. They do not offset the traffic-clogging impacts of private ride TNC services like UberX and Lyft."[24]

Schaller's findings appear to be consistent with views of how TNCs impacted San Francisco. With this actuality in mind, we see that this specific outcome could count as a distinct form of mono-technical saturation of a highly specific kind of transportation part that requires distinguishing them from other parts (e.g., private cars) that are extremely similar. Although the case of San Francisco is an isolated incident and does not represent a sufficient sample size, it nevertheless can serve as an exemplar of the kind of outcomes that other cities should aim to avoid. New York City's recent policy to limit the number of TNC drivers could be thought about as a measure to mitigate the kind of situation that emerged in San Francisco.[25]

The point here is not to vilify TNCs. While these claims challenge ideas that TNCs are a boon to cities, they did have several benefits that deserve attention. For instance, ride-sourcing companies were able to provide services to vulnerable people who lacked effective transportation.[26] TNCs were also able to make mobility feasible during late-night hours, and improved economic efficiency as well.[27] Considering that several cities' transport services must address the "last-mile" problem (getting residents from their homes to public transportation, e.g., bus or train stops), TNCs maintain that they can help connect riders with public transit.[28] Although these reasons provide ample motivation for planners to favor them when planning for the future, they should have some reservations.

For example, many municipalities began to rely on TNCs to connect transport users with transit services when dealing with the last-mile problem.[29] Cities would subsidize these companies or form partnerships.[30] Metro Transit, the transportation service found in Minneapolis and the Saint Paul region, will reimburse users up to $100 or four trips via TNC or Taxicabs.[31]

The City of Dallas' transportation authority, Dallas Area Rapid Transit (DART), has partnered with Uber to help riders connect with transport services.[32] While this move might sound like a practical solution to this problem, transit professionals such as Jarrett Walker have criticized such practices, remaining skeptical of TNCs' motivations to deliver solutions.[33] Instead of fixing existing transport systems with measures developed within the municipality, one could argue that cities that engage in such practices are merely passing a public issue along to a private company, which might not have the people's interest serving as the primary motivation. However, a municipality should not be driven solely by the same economic considerations, even though such an aspect is an area of concern.

Bearing these notions in mind, TNCs are still in their early stages, meaning that developing corrective measures to mitigate harmful effects that immediately materialize could be wise for the long term, especially considering that ride-sourcing services could evolve into operations wherein only AVs were used.[34] That is to say, becoming well informed of the problems that one could expect to find with companies that deal with the public's transport needs could serve as close precedents for future transportation technologies that might have similar effects. This point suggests that we should not shy away from TNCs making the transition from human-driver to machine-driver, but that we could facilitate such a shift to include human values and socially just avenues to urban sustainability and flourishing, along with the means to support such worthwhile goals.

In the section that follows, I show how employing the pattern of thinking behind TNCs to AVs could follow a similar, problematic course. Although these patterns are alike, the number of predictions increases with AVs, moving the inspiration behind them closer to utopian thinking than we saw with TNCs. In turn, consideration for uncertainty requires additional attention and scrutiny to determine if their promise, if unfulfilled, is worth the price that cities will pay for failed or delayed enthusiasm toward driverless vehicles. The goal is to reveal that some of the assumptions behind decisions that transport professionals have made regarding AVs' future hold the possibility of harming urban residents in some cases, and they might exacerbate damages in other instances. These points aside, the task at hand is (surprisingly) not to wholesale argue against the possibility of AVs entering our streets.

AUTOMATED VEHICLES (AVS) AND AN UNCERTAIN FUTURE

TNCs were once held as having great promise for solving traffic woes, which led to policies that limited the number of TNC vehicles on the road in some instances. This point suggests that AVs should not be thought about without limitations. In turn, it should not seem unrealistic to extend such considerations to the implementation of AVs onto our city streets. Despite having promise for improving the conditions for urban life, determining how and when AVs will deliver is a different matter. When it comes to the future of AVs, researchers note that their future has uncertainties.[35] They provide a strong reason to have reservations about viewing AVs as the transportation plan for the future of urban mobility. While such an outcome suggests that an event of this magnitude could lead to mono-technical saturation, one could make a case that it would not qualify as an area of concern if it does not produce conditions associated with instances that create additional problems for residents.

As stated earlier, there is no guarantee if or when AVs will effectively and finally solve such problems. Consider, for instance, that AV researchers, industry leaders, transportation experts, and municipal officials cannot agree when driverless vehicles will become a feasible actuality.[36][37] Having a consensus is not a necessary condition for going forward with plans to include AVs into cities' transportation systems. Still, this situation lends itself to the idea that AVs might not be available for wide-scale public consumption for quite some time. This notion suggests that if AVs are to mitigate existing harms, then our desire to address these problems hinges upon when, if, and how these technologies become urban mainstays. This fact does not mean that transportation professionals should abandon plans to include AVs as parts of transportation wholes, but it does suggest that they might want to regulate the amount of attention they receive, or at least make plans to include them as one piece of a much larger whole that consists of many other kinds of parts.

There are at least two reasons behind this view that require attention. First, such an undertaking does not escape the worrisome conditions that mono-technical saturation could produce. The idea that deserves attention is that while engineers can account for how AVs will react under controlled conditions, cities are anything but stable, almost by definition. Due to this condition, we can also anticipate that bringing these devices into our cities means that we have to accept that unforeseen circumstances will hold steady as expected with new parts of the city. Planning to expect the unexpected might not be an attainable goal or at least one held without significant reservation by the people who will be affected by AVs' presence as they navigate our streets. Second, if we imagine the costs, labor, and adjustments that would come with their implementation would be significant, the idea that we must be incredibly judicious comes to mind, considering that other social priorities could receive less support.

One could argue that AVs' potential benefits are concerns that are best left up to the scientists, engineers, and professionals who have the technical expertise to make meaningful claims. This idea suggests that urban dwellers will have to yield to their advice, but doing so could damage any attempt to build trust if such conditions fail to manifest. For instance, the reality of such benefits, as seen above, remains dubious. For the worst-case scenario, the technology ultimately fails to safely deliver a mode of personal transit that lacks the shortcomings that are commonly associated with our horseless carriages. All of our AV dreams are for nothing. The fact that driverless vehicles are currently in experimental phases suggests that the situation above is unlikely, especially considering that limited use of AVs is now available in Phoenix, Arizona, and has ample financial backing.[38]Although this progress is exciting in terms of advancing technologies, the lack of consensus on the part of the experts suggests that it could be quite a while until AVs are a permanent and ubiquitous part of cityscapes.

The problem is that transportation professionals could base their decisions to prioritize AVs on the predicted benefits listed earlier, championing this view when engaging in co-planning. It should be implied that they would be acting to achieve the desired result of moving people to and from their destinations to secure better outcomes than current models can provide. While this notion could hold when discussing any form of transportation, the issue is that the assumed view supports arguments wherein AVs could receive prioritized financial support for their alleged benefits to society and the environment.

Aside from the idea of co-planning as essential, one could easily make a cogent case that AVs could satisfy all of the requirements of moral ordering, eliminating the need to even think about the problem of moral prioritization in transportation systems. However, basing a decision on the hope that the expected results manifest in the predicted manner is not guaranteed. If the outcome diverges from the original plan or fails to materialize, then the intentions behind such decisions will not work for the solution that is required.

This element is the primary hindrance behind transportation planning for AVs, given the lack of consensus on when these new technologies will be ready for social integration, as indicated above. There is no assurance that the desired outcomes will manifest, or they could happen at a much later date than anticipated, perhaps several years or decades. Suppose these technologies do take extremely long periods to be ready for our city streets. In that case, transportation professionals could miss opportunities to improve transportation systems that are already suffering from issues such as social injustice (which need to be determined on a case-by-case basis). If they could have known that AVs were not going to work or that they would take decades to be ready for wide-scale application, then they could have financed other modes of mobility that could have supported worthwhile goals such as socially just sustainability.

While giving so much attention to AVs might signal that transport planners are adequately giving emerging technologies attention, they are prioritizing future problems that do not fully exist ahead of issues that trouble urban residents today. Such actions would count as problems of moral prioritization. This point does not suggest that planners should discount the importance of planning with emerging technologies such as AVs in mind, but they should balance such consideration with existing conditions, giving priority to the problems that already exist before addressing issues that are likely to emerge, eventually. Although this concern might seem relatively straightforward to some people, it is a commonly held attitude that guides transportation planning.[39]

Consider, for instance, that Martens argues that making these kinds of decisions align with the typical type of actions that transport planners take when determining a city's transportation future.[40] While it might seem wise to approach planning in this fashion, the problem with this approach is that preexisting issues that concern justice do not receive attention.[41] Harms that stem from ill-functioning and unjust transportation systems can continue, and the people living with such burdens do not receive much relief.[42] That is to say, if there is a discrepancy in the distribution of services, an issue that raises concerns for transportation justice, problems could persist indefinitely or worsen.[43]

Keeping this point in mind, it seems fitting to think that we should abandon the view that AVs can solve all of our transportation problems, making them the center of our thinking on the future of urban transportation. We should not be serving driverless vehicles. They should help us. This idea means that we should focus our efforts on ways to solve problems that currently exist—while keeping a keen eye on the future. If AVs can provide relief to such problems, then there is no good reason why we should not use driverless vehicles (not just cars but also some buses) to help remedy such situations. Still, there is also no good reason why cities cannot plan for a future with AVs, thinking through the technology *with* urban travelers and dwellers.

This idea brings urban enlightenment into view, showing how transportation experts can help create opportunities for residents to gain access to the experiences that would facilitate their right to shape the city. In turn, thinking about AVs in this manner shows how they can be devices that work toward goals such as urban sustainability and transportation justice, rather than just supporting AVs while hoping that they deliver the same results. If we pair this thinking with the with leading experts who work on mobility systems, such ideas are often compatible with some of their ideological practices.

For example, Jarret Walker holds that transport planners should not be advocates for any particular mode of mobility, but instead, they focus on the task of human transit.[44] This idea underscores the importance of avoiding placing bets on predictions to come true when it comes to the future of transportation systems and driverless vehicles. In turn, discussions about AVs should focus on how or if they should fit in with existing transportation systems, not the opposite. Consider, for instance, that this notion remains consistent with approaches to transportation planning and engineering that champion a multi-modal approach. The idea behind such a course of action would employ modes of transportation that are best suited for the task. This approach is inherently resistant to dogmatic allegiances toward mobility modes, such as horseless carriages and driverless vehicles. Yet, engaging in such a practice is sure to raise additional concerns.

Here are two such issues that are worth mentioning. If we recall the problem of mono-technical saturation, having a city with AVs as the primary mode of mobility could lead to unanticipated outcomes. People would be stuck with issues that emerge from the overreliance on AVs. While such outcomes are not inherently bad, removing these parts could be challenging once they became part of the whole transportation system. Yet, the significant issue is that if people became reliant on these parts and they became the only option, then people would not have access to other kinds of parts that might produce better outcomes. Cities would move from forced-car ownership to forced-AV usage wherein there is little room for an alternative. Due to this situation, neglected or discounted parts could include bicycles that could improve public health or bus lines that can promote feelings of community or well-being, along with the feelings of familiarity for seeing the same travelers daily.

Although we cannot change the past, we can co-plan for the future, which at least embodies a view that respect is desired in a manner that can emerge. In turn, such an attitude can not only provide the kind of insights that can attend to issues concerning the introduction of new transportation parts into larger systems, but they also restructure the power dynamics of mobility affairs while establishing the conditions for moving toward urban enlightenment.

This idea emblematizes that going against established transportation practices could benefit emerging issues in urban mobility, a needed measure that can help us shape the transportation systems that will in turn influence our lives. Considering that this notion will hold, the task of providing urban dwellers with a mobility system that is built for them has no end in sight. While it takes engineers and planners to make the trains run and to have them arrive and depart on schedule, the process of moral ordering, as one technology among others, must hold steady to ensure that history does not repeat itself but is made. In turn, the role of philosophy must go beyond examining cities. It must guide the process of changing them.

NOTES

1. For more information, see Austin Brown *et al.*, "An analysis of possible energy impacts of automated vehicles," in *Road vehicle automation*, eds. Gereon Meyer and Sven Beiker (Cham: Springer, 2014), 137–53. Also, see Warren Walkerand *et al.*, "Dynamic adaptive policymaking for the sustainable city: The case of automated taxis," *International Journal of Transportation Science and Technology* 6, no. 1 (2017): 1.

2. Eva Fraedrich *et al.*, "Transition pathways to fully automated driving and its implications for the sociotechnical system of automobility," *European Journal of Futures Research* 3, no. 1 (2015): 1.

3. For an example of this kind of work, see Markus Maurer *et al.*, *Autonomous driving: Technical, legal and social aspects* (Berlin Heidelberg: Springer, 2016).

4. Hans Jonas, *The imperative of responsibility: In search of an ethics for the technological age* (Chicago: University of Chicago Press, 1984), 189.

5. For example, see Patrick Lin, "Why ethics matters for autonomous cars," in *Autonomous driving*, eds. Markus Maurer, Christian Gerdes, Barbara Lenz and Hermann Winner (Berlin, Heidelberg: Springer, 2016), 69–85 Also, see Sven Nyholm *et al.*, "The ethics of accident-algorithms for self-driving cars: An applied trolley problem?," *Ethical theory and moral practice* 19, no. 5 (2016): 1275–89.

6. This notion extends to several disciplines, suggesting that one reason why researchers are exploring the challenges that AVs will bring is because the positives of these technologies outweigh the possible negatives. This notion is also consistent with a Hearing for the Subcommittee on Highways and Transit of the Committee on Transportation and Infrastructure, House of Representatives, One Hundred and Thirteenth Congress, First Session, November 19, 2013. They acknowledge that the introduction of AVs will bring several challenges to society, but that these vehicles will improve roadway safety, decrease traffic, reduce auto emissions, and bolster the workforce. They show that driverless vehicles have support at all levels of government, which could someday deliver a mobility revolution in the United States. For more information, see Subcommittee on Highways and Transit; Committee on Transportation and Infrastructure. House. How Autonomous Vehicles Will Shape the Future of Surface Transportation, 2013. https://www.govinfo.gov/content/pkg/CHRG-113hhrg85609/pdf/CHRG-113hhrg85609.pdf.

7. Wanis Kabbaj's Ted Talk, "What a driverless world could look like," is a good example. https://www.youtube.com/watch?v=OlLFK8oSNEM.

8. While industry representatives typically present such views, there are a few scholars in academia whose work supports and or leans in the direction of the assumed view. For example, see Greenblatt and Shaheen, *Automated vehicles*, 74–81.

9. Brown, Gonder *et al.*, "Analysis," 137ff.

10. Ibid.

11. William Morrow *et al.*, "Key factors influencing autonomous vehicles' energy and environmental outcome," in *Road vehicle automation*, eds. Gereon Meyer and Sven Beiker (Cham: Springer, 2014), 127–35.

12. Taylor Stone *et al.*, "Driving in the dark: Designing autonomous vehicles for reducing light pollution," *Science and Engineering Ethics* 26, no. 1 (2020): 388.

13. Hubert Iglińskia *et al.*, "Analysis of the potential of autonomous vehicles in reducing the emissions of greenhouse gases in road transport," *Procedia Engineering* 192 (2017): 357.

14. Armin Grunwald, "Societal risk constellations for autonomous driving: Analysis, historical context and assessment," in *Autonomous driving*, eds. Markus Maurer, Christian Gerdes, Barbara Lenz and Hermann Winner (Berlin, Heidelberg: Springer, 2016), 650, 641–63.

15. Through focusing on the (collective) positive appeal of AVs, I do not mean to discount the possible challenges that driverless vehicles will present. The underscored point here is that AV's benefits make it worthwhile to develop solutions and workarounds to such challenges.

16. Shelley Kagan, *Normative ethics* (New York City: Westview Press, 1998), 64. Also, I am aware that I am criticizing consequentialist views while, essentially, also making a consequentialist argument. However, not all consequentialist arguments are the same. They simply share co-extensive qualities that pertain to their structure, and my focus on outcomes accepts that we can make adjustments to achieve outcomes that either fail or do not materialize. In turn, if, as in the case of adding, taking away, or rearranging parts of transportation system, if they fail to deliver, we can still make adjustments that mitigate such effects.

17. Jonas, *Imperative*, 203. He addresses the charge of being against technology, holding that we should not entirely abandon technological pursuits. The spirit of the present critique of driverless vehicles should not be construed as a form of technophobia. In turn, I am also not accusing researchers or developers of having technophilia. For more information on these topics, see Alan Drengson, "Four philosophies of technology," in *Technology and values: Essential readings*, ed. Craig Hanks (West Sussex, UK), 26–37.

18. Lisa Rayle *et al.*, *App-based, on-demand ride services: Comparing taxi and ridesourcing trips and user characteristics in San Francisco* (No. UCTC-FR-2014-08) (Berkeley, CA: University of California Transportation Center, 2014), 1.

19. Uber, "What is Uber pass?," https://help.uber.com/riders/article/what-is-uber-pass?nodeId=1f9dbe59-8c13-45f6-8048-689e9c76b3ac.

20. Farhad Manjoo, "With Uber, less reason to own a car," *New York Times*, June 12, 2014. https://www.nytimes.com/2014/06/12/technology/personaltech/with-ubers-cars-maybe-we-dont-need-our-own.html.

21. Sneha Roy, *Quantifying the impact of transportation network companies (TNCs) on traffic congestion in San Francisco* (Lexington, KY: University of Kentucky, 2019), 108.

22. Gregory Erhardt, *et al.*, "Do transportation network companies decrease or increase congestion?" *Science Advances* 5, no. 5 (2019): 1–10.

23. John Barrios *et al.*, "The cost of convenience: Ridesharing and traffic fatalities," *University of Chicago, Becker Friedman Institute for Economics Working Paper* (2020): 25.

24. Bruce Schaller, *The new automobility: Lyft, Uber and the future of American cities*, 2018, 2. http://www.schallerconsult.com/rideservices/automobility.pdf.

25. It is worth mentioning that Uber has filed a lawsuit against New York City for their cap on on-demand drivers. For more information, see Dana Rubenstein, "Uber and Lyft stop accepting new drivers in New York City," *Politico*, April 29, 2019. https://www.politico.com/states/new-york/albany/story/2019/04/29/uber-and-lyft-have-stopped-accepting-new-drivers-in-new-york-city-993270.

26. Scarlett Jin *et al.*, "Ridesourcing, the sharing economy, and the future of cities," *Cities* 76 (2018): 97.

27. Jin, Kong *et al.*, "Ridesourcing," 97.

28. Uber, "Partnering with transit systems," *Supporting Cities*, 2019. https://www.uber.com/us/en/community/supporting-cities/transit/.

29. Susan Shaheen and Nelson Chan, "Mobility and the sharing economy: Potential to facilitate the first-and last-mile public transit connections," *Built Environment* 42, no. 4 (2016): 573.

30. Shaheen and Chan, "Mobility," 588.

31. Metro Transit. Guaranteed Ride Home. *Metro Transit*, 2020. https://www.metrotransit.org/guaranteed-ride-home.

32. Morgan Lyons *et al.*, "DART, Uber stepping up "complete trip" efforts," *DART News Release*, 2015. https://www.dart.org/news/news.asp?ID=1179.

33. Jarrett Walker, "Do Uber and Lyft want to connect to transit?" in *Human transit: The professional blog of public transit consultant Jarrett Walker*, 2019. https://humantransit.org/2019/04/do-uber-and-lyft-want-to-connect-to-transit.html.

34. Jeffery Greenblatt *et al.*, "Automated vehicles, on-demand mobility, and environmental impacts," *Current Sustainable/Renewable Energy Reports* 2, no. 3 (2015): 79.

35. Brown *et al.*, "Analysis," 137ff.

36. Greenblatt *et al.*, "Automated vehicles," 74ff.

37. Jeffery Greenblatt *et al.*, show that there is significant disagreement on when AVs will be a ubiquitous on roadways: "All manufacturers that have announced plans for AVs already offer or plan to release vehicles with some automated features by 2017, and level 3 systems are expected by 2017 to 2020. As mentioned above, Google has announced plans to release a level 4 system by 2017, and Tesla has announced its intention by 2020. Researchers disagree on when AVs will become generally available. IHS Automotive projects level 3 functionality by 2020, level 4 by 2025, and level 5 by 2030, with AVs reaching 9% of sales in 2035 and 90% of the vehicle fleet by 2055. Navigant Consulting was even more optimistic, expecting 75% of light-duty vehicle sales to be automated by 2035, whereas the Insurance Information Institute claims that all cars may be automated by 2030." Citations for the above estimations are in the original article by Jeffery Greenblatt *et al.* To review these references, see Jeffery Greenblatt *et al.*, "Automated vehicles, on-demand mobility, and environmental impacts," *Current Sustainable/Renewable Energy Reports* 2, no. 3 (2015): 74–81.

38. Joseph White, "Waymo opens driverless robo-taxi service to the public in Phoenix," *Reuters*, October 8, 2020. https://www.reuters.com/article/us-waymo

-autonomous-phoenix/waymo-opens-driverless-robo-taxi-service-to-the-public-in-phoenix-idUSKBN26T2Y3.

39. Karel Martens, *Transport justice: Designing fair transportation systems* (Abingdon: Routledge, 2016), 28–9.

40. Martens, *Transport*, 28–9.

41. Ibid.

42. Ibid.

43. Ibid.

44. Jarrett Walker, "Streetcars: An inconvenient truth," *Human transit: The professional blog of public transit consultant Jarrett Walker*, 2009. http://www.humantransit.org/2009/07/streetcars-an-inconvenienttruth.html.

Bibliography

Agrawal, Asha. "What do Americans think about public transit? A review of U.S. public opinion polling survey questions." *Mineta Transportation Institute Publications*, 2015.

Attfield, Robin. "Beyond anthropocentrism." *Royal Institute of Philosophy Supplements* 69 (2011): 29–46.

Alexander, Larry and Michael Moore. "Deontological Ethics," *The Stanford Encyclopedia of Philosophy* (Winter 2020 Edition). Edited by. Edward Zalta. https://plato.stanford.edu/archives/win2020/entries/ethics-deontological/.

Bagloee, Saeed Asadi, Madjid Tavana, Mohsen Asadi and Tracey Oliver. "Autonomous vehicles: Challenges, opportunities, and future implications for transportation policies." *Journal of Modern Transportation* 24, no. 4 (2016): 284–303.

Bai, Xuemei and Heinz Schandl. "Urban ecology and industrial ecology." In *Handbook of urban ecology*, edited by D. Ian, D. Goode, M. Houck and R. Wang (2010), 26–37.

Ball, Morgan and Mark Lyons. "DART, Uber stepping up 'complete trip' efforts." *DART News Release*, 2018. https://www.dart.org/news/news.asp?ID=1179.

Ballet, Jérôme and Damien Bazin. "Hans Jonas: Bridging the gap between environmental justice and environmental ethics." *Environmental Ethics* 39, no. 2 (2017): 175–91.

Barrios, John, Yael Hochberg and Livia Yi. *The cost of convenience: Ridesharing and traffic fatalities*. University of Chicago, Becker Friedman Institute for Economics Working Paper, 2018. https://research.chicagobooth.edu/-/media/research/stigler/pdfs/workingpapers/27thecostofconvenience.pdf?la=en&hash=A15B1513F98D7A17B9E37F78DD2EBDC4C6338BFA.

Beaven, Stephen. "New Portland apartment buildings with no parking have neighbors worried about congested streets." *The Oregonian*, July 12, 2012. https://www.oregonlive.com/portland/2012/07/new_portland_apartment_buildin.html.

Bettencourt, Luis and Geoffrey West. "A unified theory of urban living." *Nature* 467, no. 7318 (2010): 912–13.

Blumenberg, Evelyn and Asha Weinstein Agrawal. "Getting around when you're just getting by: Transportation survival strategies of the poor." *Journal of Poverty* 18, no. 4 (2014): 355–78.

Boonin, David. *The non-identity problem and the ethics of future people*. New York: Oxford University Press, 2014.

Borenstein, Jason, Joseph Herkert and Keith Miller. "Self-driving cars: Ethical responsibilities of design engineers." *IEEE Technology and Society Magazine* 36, no. 2 (2017): 67–75.

Bouchard, Mikayla. "Transportation emerges as crucial to escaping poverty." *The New York Times*, May 7, 2015. http://www.nytimes.com/2015/05/07/upshot/transportation-emerges-as-crucial-to-escaping-poverty.html.

Brey, Philip. "From moral agents to moral factors: The structural ethics approach." In *The moral status of technical artefacts*, edited by Peter Kroes and Peter-Pauk Verbeek, 125–42. Dordrecht, NL: Springer, 2014.

Brown, Austin, Jeffrey Gonder and Brittany Repac. "An analysis of possible energy impacts of automated vehicles." In *Road vehicle automation*, edited by Gereon Meyer and Sven Beiker, 137–53. Cham: Springer, 2014.

Callicott, J. Baird. *In defense of the land ethic: Essays in environmental philosophy*. Albany, NY: SUNY Press, 1989.

Callicott, J. Baird. *Beyond the land ethic: More essays in environmental philosophy*. Albany, NY: SUNY Press, 1999.

Caro, Robert. *The power broker: Robert Moses and the fall of New York*. New York: Alfred A Knopf Incorporated, 1974.

Caso, Antonio. "The human person and the totalitarian state." In *Latin American philosophy in the twentieth century*, edited by Jorge Gracia, 43–7. Amherst, NY: Prometheus Books, 1986.

City of Portland, Office of Transportation. *Bicycle master plan*. https://www.portlandoregon.gov/transportation/article/369990. Last updated, July 1, 1998.

City of Seattle. "6 case studies in urban freeway removal." *Seattle urban mobility plan*. January 2008. https://www.cnu.org/sites/default/files/Nashville%20Case%20Study%202%20-%20Seattle.pdf.

Cornwell, Erin and Linda Waite. "Social disconnectedness, perceived isolation, and health among older adults." *Journal of Health and Social Behavior* 50, no. 1 (2009): 31–48.

de Vignemont, Frederique, Manos Tsakiris and Patrick Haggard. "Body mereology." In *Human body perception from the inside out*, edited by Gunther Knoblich, Ian Thornton, Marc Grosjean and Maggie Shiffrar. New York, NY: Oxford University Press, 148–70.

Doeden, Matt. *Choppers*. Minneapolis, MN: Lerner Publications Co., 2008.

Drengson, Alan. "Four philosophies of technology." In *Technology and values: Essential readings*, edited by Craig Hanks, 26–37. West Sussex: UK.

Dussel, Enrique. *Philosophy of liberation*. Translated by Aquila Martinez and Christine Morkovsky. Maryknoll, NY: Orbis Books, 1985.

Epting, Shane. "Questioning technology's role in environmental ethics: Weak anthropocentrism revisited." *Interdisciplinary Environmental Review* 11, no. 1 (2010): 18–26.

Epting, Shane. "On moral prioritization in environmental ethics: Weak anthropocentrism for the city." *Environmental Ethics* 39, no. 2 (2017): 131–46.

Epting, Shane. "The Peñalosa Principle of Transportation Democracy: lessons from Bogotá on the morality of urban mobility." *Science and Engineering Ethics* 23, no. 4 (2017): 1085-1096.

Epting, Shane. "Automated vehicles and transportation justice." *Philosophy & Technology* 32, no. 3 (2019): 389–403.

Epting, Shane. "Urban infrastructure and the problem of moral praise." *Techné: Research in Philosophy and Technology* 25, no. 2 (2021): forthcoming.

Epting, Shane. "On municipalities as technologies." *Philosophy & Technology* (2021): 1–11. https://doi.org/10.1007/s13347-020-00438-z.

Epting, Shane. "Ethical requirements for transport systems with automated buses." *Technology in Society* 64 (2021): 101508.

Erhardt, Gregory, Sneha Roy, Drew Cooper, Bhargava Sana, Mei Chen and Joe Castiglione. "Do transportation network companies decrease or increase congestion?." *Science Advances* 5, no. 5 (2019): eaau2670.

Eugensson, Anders, Mattias Brännström, Doug Frasher, Marcus Rothoff, Stefan Solyom and Alexander Robertsson. "Environmental, safety legal and societal implications of autonomous driving systems." In *International Technical Conference on the Enhanced Safety of Vehicles (ESV). Seoul, South Korea*, vol. 334. 2013. Technical Paper, no. 13-0467.

Ezike, Richard, Peter Tatian and Gabriella Velasco. "Defining "communities of concern." *Transportation Planning: A Review of How Planners Identify Underserved Communities*." (2020): 1–12. https://www.urban.org/sites/default/files/publication/102746/defining-communities-of-concern-in-transportation-planning_1.pdf.

Ferré, Frederick. "Personalistic organicism: Paradox or paradigm?." *Royal Institute of Philosophy Supplements* 36 (1994): 59–73.

Figueroa, Robert. "Evaluating environmental justice claims." In *Forging environmentalism: Justice, livelihood, and contested environments*, edited by Joan Bauer, 360–76. Amonk, NY: M.E. Sharpe, 2006.

Forsyth, Ann and Michael Southworth. "Cities afoot—Pedestrians, walkability and urban design." *Journal of Urban Design* 13, no. 1 (2008): 1–3.

Foth, Marcus. *Handbook of research on urban informatics: The practice and promise of the real-time city*. Hershey, PA: Information Science Reference, 2009.

Fraedrich, Eva, Sven Beiker and Barbara Lenz. "Transition pathways to fully automated driving and its implications for the sociotechnical system of automobility." *European Journal of Futures Research* 3, no. 1 (2015): 1–11.

Frank, Lawrence D., James F. Sallis, Terry L. Conway, James E. Chapman, Brian E. Saelens and William Bachman. "Many pathways from land use to health: Associations between neighborhood walkability and active transportation, body mass index, and air quality." *Journal of the American Planning Association* 72, no. 1 (2006): 75–87.

Gerdes, Christian and Sarah M. Thornton. "Implementable ethics for autonomous vehicles." In *Autonomous driving*, edited by Markus Maurer, Christian Gerdes, Barbara Lenz and Hermann Winner, 87–102. Berlin, Heidelberg: Springer, 2016.

Gini, Al and Alexei Marcoux. *The ethics of business: A concise introduction.* Lanham, MD: Rowman & Littlefield Publishers, 2011.

Glaeser, Edward. *Triumph of the city: How our greatest invention makes us richer, smarter, greener, healthier and happier*. New York City: Penguin, 2011.

Gordon, Charles. "Automobiles vs. railways for urban transportation." *Electrical Engineering* 50, no. 1 (1931): 4–8.

Gordon, Lewis. *Disciplinary decadence: Living thought in trying times*. New York: Routledge, 2015.

Grabar, Henry. "They can just take an Uber." *Slate*, December 14, 2016. http://www.slate.com/articles/business/metropolis/2016/12/cities_are_cutting_ https://escholarship.org/uc/item/3577d04p.

Gracia, Jorge. "Antonio Caso." In *Latin American philosophy in the twentieth century*, edited by Jorge Gracia, 41–2, Amherst, NY: Prometheus Books, 1986.

Greenblatt, Jeffery and Susan Shaheen. "Automated vehicles, on-demand mobility, and environmental impacts." *Current Sustainable/Renewable Energy Reports* 2, no. 3 (2015): 74–81.

Grunwald, Armin. "Societal risk constellations for autonomous driving: Analysis, historical context and assessment." In *Autonomous driving*, edited by Markus Maurer, Christian Gerdes, Barbara Lenz and Hermann Winner, 641–63. Berlin, Heidelberg: Springer, 2016.

Guha, Ramachandra. "Radical American environmentalism and wilderness perservation: A third world critique." *Environmental Ethics* 11, no. 1 (1989): 71–83.

Handy, Susan. "Making US cities pedestrian-and bicycle-friendly." In *Transportation, land use, and environmental planning*, edited by Elizabeth Deakin, 169–87. Elsevier, 2020.

Hargrove, Eugene. "Weak anthropocentric intrinsic value." *The Monist* 75, no. 2 (1992): 183–207.

Hargrove, Eugene. "Forward." In *Faces of environmental racism: Confronting issues of global justice*, edited by Laura Westra and Bill Lawson. Lanham, MD: Rowman & Littlefield Publishers, 2001.

Harvey, David. "The right to the city." *New Left Review* 53 (2008): 23–40.

Heffner, Reid, Tom Turrentine and Kenneth S. Kurani. *A primer on automobile semiotics*. Davis, CA: Institute of Transportation Studies, University of California, 2006.

Hennig, Ernest, Christian Schwick, Tomáš Soukup, Erika Orlitová, Felix Kienast and Jochen Jaeger. "Multi-scale analysis of urban sprawl in Europe: Towards a European de-sprawling strategy." *Land Use Policy* 49 (2015): 483–98.

Higgins, Tim. "Driverless cars tap the brakes after years of Hype; developers take a more cautious, low-key approach in testing and talking about autonomous vehicles after Uber crash." *The Wall Street Journal*, January 17, 2019. https://www.wsj.com/articles/driverless-cars-tap-the-brakes-after-years-of-hype-11547737205.

Huang, Chun-Wei, Robert I. McDonald and Karen C. Seto. "The importance of land governance for biodiversity conservation in an era of global urban expansion." *Landscape and Urban Planning* 173 (2018): 44–50.

Igliński, Hubert and Maciej Babiak. "Analysis of the potential of autonomous vehicles in reducing the emissions of greenhouse gases in road transport." *Procedia Engineering* 192 (2017): 353–8.

J.D. Power. "Brand loyalty increasing among new-vehicle buyers, J.D. Power Finds." *J.D. Power Press Release*, July 15, 2020. https://www.jdpower.com/sites/default/files/file/202007/2020046%20U.S.%20Automotive%20Brand%20Loyalty.pdf.

James, Leon. *Road rage and aggressive driving: Steering clear of highway warfare.* Amherst, NY: Prometheus Books, 2009.

Jin, Scarlett, Hui Kong, Rachel Wu and Daniel Z. Sui. "Ridesourcing, the sharing economy, and the future of cities." *Cities* 76 (2018): 96–104.

Jonas, Hans. *The imperative of responsibility: In search of an ethics for the technological age.* Chicago: University of Chicago Press, 1984.

Kagan, Shelley. *Normative ethics.* New York City: Westview Press, 1998.

Kant, Immanuel. "Answering the question: What is enlightenment?" *Berlin Monthly.* Berlin: Berlin Monthly, 1784.

King Jr., Martin Luther. "Letter from Birmingham jail." In *Liberating faith: Religious voices for justice, peace, & ecological wisdom*, edited by Roger Gottieb, 177–87, Lanham, MD: Rowman & Littlefield Publishers, 2003.

Kirkman, Robert. *The ethics of metropolitan growth: The future of our built environment.* London: Bloomsbury Publishing, 2010.

Kollies Parts. "What's a bobber and a chopper?" 2021. https://www.kolliesparts.nl/en/blogs/kollies-parts-blog/whats-a-bobber-and-whats-a-chopper/.

Kroll, Gary. "An environmental history of roadkill: Road ecology and the making of the permeable highway." *Environmental History* 20, no. 1 (2015): 4–28.

Kwan, Soo Chen and Jamal Hisham Hashim. "A review on co-benefits of mass public transportation in climate change mitigation." *Sustainable Cities and Society* 22 (2016): 11–18.

Lando, Giorgio. *Mereology: A philosophical introduction.* London and New York City: Bloomsbury Publishing, 2017.

Latour, Bruno. "Where are the missing masses?, The sociology of a few mundane artifacts." In *Shaping technology, building society: Studies in sociotechnological change*, edited by Wiebe Bjiker and John Law, 225–58. Cambridge, MA: MIT Press, 1992.

Lawson, Bill. "Living for the city: Urban United States and environmental justice." In *Faces of environmental racism: Confronting issues of global justice*, edited by Laura Westra and Bill Lawson, 41–55. Lanham, MD: Rowman & Littlefield Publishers, 1995.

Leigh-Hunt, Nicholas, David Bagguley, Kristin Bash, Victoria Turner, Stephen Turnbull, N. Valtorta and Woody Caan. "An overview of systematic reviews on the public health consequences of social isolation and loneliness." *Public Health* 152 (2017): 157–71.

Leopold, Aldo. *A Sand County Almanac and sketches here and there.* New York: Oxford University Press, 1949.

Le Pira, Michela, Giuseppe Inturri and Matteo Ignaccolo. "Competence, interest and power in participatory transport planning: Framing stakeholders in the "participation cube"." *Transportation Research Procedia* 48 (2020): 2385–400.

Levinas, Emmanuel. *Totality and infinity: An essay on exteriority* [French original 1961]. Translated by Alphonso Lingis. Pittsburg: Duquesne University Press, 1969.

Lin, Patrick. "Why ethics matters for autonomous cars." In *Autonomous driving*, edited by Markus Maurer, Christian Gerdes, Barbara Lenz and Hermann Winner, 69–85. Berlin, Heidelberg: Springer, 2016.

Lubitow, Amy, JaDee Carathers, Maura Kelly and Miriam Abelson. "Transmobilities: Mobility, harassment, and violence experienced by transgender and gender nonconforming public transit riders in Portland, Oregon." *Gender, Place & Culture* 24, no. 10 (2017): 1398–1418.

Maantay, Juliana. "Asthma and air pollution in the Bronx: Methodological and data considerations in using GIS for environmental justice and health research." *Health & Place* 13, no. 1 (2007): 32–56.

Mahmassani, Hani, Meead Saberi and Ali Zockaie. "Urban network gridlock: Theory, characteristics, and dynamics." *Transportation Research Part C: Emerging Technologies* 36 (2013): 480–97.

Manjoo, Farhad. "With Uber, less reason to own a car." *New York Times*, June 12, 2014. https://www.nytimes.com/2014/06/12/technology/personaltech/with-ubers-cars-maybe-we-dont-need-our-own.html.

Martens, Karel. *Transport justice: Designing fair transportation systems*. Abingdon: Routledge, 2016.

Mattioli, Giulio. "'Forced car ownership' in the UK and Germany: Socio-spatial patterns and potential economic stress impacts." *Social Inclusion* 5, no. 4 (2017): 147–60.

Maurer, Markus, Christian Gerdes, Barbara Lenz and Hermann Winner. *Autonomous driving: Technical, legal and social aspects*. Berlin Heidelberg: Springer, 2016.

Metro Transit. Guaranteed ride home. *Metro Transit*, 2020. https://www.metrotransit.org/guaranteed-ride-home.

McCarthy, Timothy. "A note on unrestricted composition." *Thought: A Journal of Philosophy* 4, no. 3 (2015): 202–11.

Milam, Ronald, Marc Birnbaum, Chris Ganson, Susan Handy and Jerry Walters. "Closing the induced vehicle travel gap between research and practice." *Transportation Research Record* 2653, no. 1 (2017): 10–16.

Mills, Charles. "Black trash." In *Faces of environmental racism: Confronting issues of global justice*, edited by Laura Westra and Bill Lawson, 73–94. Lanham, MD: Rowman & Littlefield Publishers, 2001.

Minteer, Ben. "Anthropocentrism." In *Encyclopedia of environmental ethics and philosophy*, edited by Baird Callicott and Robert Frodeman. Farmington Hills, MI: Macmillan Reference USA/Gale Cengage Learning, 2009.

Mishra, Sabyasachee, Timothy Welch and Manoj Jha. "Performance indicators for public transit connectivity in multi-modal transportation networks." *Transportation Research Part A: Policy and Practice* 46, no. 7 (2012): 1066–85.

Morrow, William, Jeffery Greenblatt, Andrew Sturges, Samveg Saxena, Anand Gopal, Dev Millstein, Nihar Shah and Elisabeth Gilmore. "Key factors influencing autonomous vehicles' energy and environmental outcome." In *Road vehicle automation*, edited by Gereon Meyer and Sven Beiker, 127–35. Cham: Springer, 2014.

Nagel, Thomas. "What is it like to be a Bat?." *The Philosophical Review* 83, no. 4 (1974): 435–50.

Nash, Jeff. "Bus riding: Community on wheels." *Urban Life* 4, no. 1 (1975): 99–124.

National Public Radio. (2019). Why phoenix area residents are attacking Waymo's self-driving fleet. *All things considered.* https://www.npr.org/2019/01/02/681752256/why-phoenix-area-residents-are-attacking-waymos-self-driving-fleet.

Nieuwenhuijsen, Mark and Haneen Khreis. "Car free cities: Pathway to healthy urban living." *Environment International* 94 (2016): 251–62.

Noll, Samantha and Laci Hubbard-Mattix. "Health justice in the city: Why an intersectional analysis of transportation matters for bioethics." *Essays in Philosophy* 20, no. 2 (2019): 130–45.

Nordström, Maria, Sven Ove Hansson and Muriel Beser Hugosson. "Let me save you some time ... on valuing travelers' time in urban transportation." *Essays in Philosophy* 20, no. 2 (2019): 206–29.

Norton, Bryan. "Environmental ethics and weak anthropocentrism." *Environmental Ethics* 6, no. 2 (1984): 131–48.

Obama, Barack. "Weekly address: Protecting our planet for future generations." https://obamawhitehouse.archives.gov/the-press-office/2015/10/24/weekly-address-protecting-our-planet-future-generations.

OPAL Environmental Justice Portland. *Adela's journey*, 2012. https://www.youtube.com/watch?v=IjxebYmey44.

Payne, Briana. "Oral history of Bonton and ideal neighborhoods in Dallas, Texas." MA Thesis, University of North Texas, 2015.

Pelzer, Peter. "Bicycling as a way of life: A comparative case study of bicycle culture in Portland, OR and Amsterdam." In *7th Cycling and Society Symposium* (2010): 1–13.

Peñalosa, Enrique. "Why buses represent democracy in action." *TedCity2.0*, September 2013. https://www.ted.com/talks/enrique_penalosa_why_buses_represent_democracy_in_action#t-51086.

Pineda, Andrés Felipe Valderrama. "What can engineering systems teach us about social (In) justices? The case of public transportation systems." In *Engineering education for social justice*, edited by Juan Lucena, 203–26. Springer, Dordrecht, 2013.

Rawls, John. *A theory of justice.* Cambridge, MA: Harvard University Press, 1999.

Rayle, Lisa, Susan Shaheen, Nelson Chan, Danielle Dai and Robert Cervero. *App-based, on-demand ride services: Comparing taxi and ridesourcing trips and user characteristics in San Francisco* (No. UCTC-FR-2014-08). Berkeley, CA: University of California Transportation Center, 2014.

Reed, Trevor. *INRIX Global Traffic Scorecard*, 2019. https://inrix.com/scorecard/.

Rolston, Holmes. *Environmental ethics: Duties to and values in the natural world.* Philadelphia: Temple University Press, 1988.

Rolston, Holmes. *Conserving natural value.* New York City: Columbia University Press, 1994.

Routley, Richard and Val Routley. "Nuclear energy and obligations to the future." *Inquiry* 21, no. 1–4 (1978): 133–79.

Roy, Sneha. *Quantifying the impact of transportation network Companies (TNCs) on Traffic Congestion in San Francisco.* Unpublished doctoral dissertation, University of Kentucky, Lexington, Kentucky, 2019.

Rubenstein, Dana. "Uber and Lyft stop accepting new drivers in New York City." *Politico*, April 29, 2019. https://www.politico.com/states/new-york/albany/story/2019/04/29/uber-and-lyft-have-stopped-accepting-new-drivers-in-new-york-city-993270.

Russell, George. "Irish agricultural organisation society." Notes of the Week: Fair Play in Legislation. *The Irish Homestead: The Organ of Irish Agricultural and Industrial Development*, 17 no. 53 (1910): 1086–90.

Rytwinski, Trina, Kylie Soanes, Jochen AG Jaeger, Lenore Fahrig, C. Scott Findlay, Jeff Houlahan, Rodney Van Der Ree and Edgar A. van der Grift. "How effective is road mitigation at reducing road-kill? A meta-analysis." *PLoS One* 11, no. 11 (2016): e0166941.

Sansone, Randy and Lori Sansone. "Road rage: What's driving it?" *Psychiatry* 7, no. 7 (2010): 14–18.

Sartre, Jean-Paul. *Existentialism.* Translated by Bernard Frechtman. New York: Philosophical Library, 1947.

Schaller, Bruce. *The new automobility: Lyft, Uber and the future of American cities.* 2018. http://www.schallerconsult.com/rideservices/automobility.pdf.

Schneider, Robert, Jason Vargo and Aida Sanatizadeh. "Comparison of US metropolitan region pedestrian and bicyclist fatality rates." *Accident Analysis & Prevention* 106 (2017): 82–98.

Scott, Aaron. "By the grace of god." *Portland Monthly.* March 2012. https://www.pdxmonthly.com/articles/2012/2/17/african-american-churches-north-portland-march-2012.

Seiler, Andreas and J. O. Helldin. "Mortality in wildlife due to transportation." In *The ecology of transportation: Managing mobility for the environment*, edited by John Davenport and Julia Davenport, 165–89. Dordrecht, NL: Springer, 2006.

Shaheen, Susan and Nelson Chan. "Mobility and the sharing economy: Potential to facilitate the first-and last-mile public transit connections." *Built Environment* 42, no. 4 (2016): 573–88.

Simons, Peter. "Varieties of parthood: Ontology learns from engineering." In *Philosophy and engineering: Reflections on practice, principles and process*, edited by Diane Michelfelder, Natasha McCarthy, David E. Goldberg, 151–63. Dordrecht: Springer, 2013.

Sims, Ralph, Roberto Schaeffer, F. Creutzig, X. Cruz-Núñez, M. D'agosto, D. Dimitriu, M. J. Figueroa Meza et al. "Transport climate change 2014: Mitigation of climate change." In *Contribution of working group III to the fifth assessment report of the intergovernmental panel on climate change*, edited by O Edenhofer et al., 599–670. Cambridge and New York: Cambridge University Press, 2014. http://www.ipcc.ch/pdf/assessment-report/ar5/wg3/ipcc_wg3_ar5_chapter8.pdf.

Singleton, Patrick A. "How useful is travel-based multitasking? Evidence from commuters in Portland, Oregon." *Transportation Research Record* 2672, no. 50 (2018): 11–22.

Steiner, Achim. "Forward." *United Nations Environment Programme* (2013): 4–5. https://wedocs.unep.org/bitstream/handle/20.500.11822/8488/CityLevel_FReport_EN.pdf?sequence=1&isAllowed=y.

Swilling, Mark, Hajer, Maarten, Baynes, Maarten, Bergesen, Joe, Labbé, Françoise, Musango, Josephine, Ramaswami, Anu, Robinson, Blake, Salat, Serge, Suh, Sangwon, Currie, Paul,Fang, Andrew, Hanson, Aaron, Kruit, Katja, Reiner, Mark, Smit, Suzanne and Samuel Tabory. "The weight of cities: Resource requirements of future urbanization." *International Resource Panel* (2018). http://wedocs.unep.org/bitstream/handle/20.500.11822/31623/TWOC1.pdf?sequence=1&isAllowed=y.

Tajalli, Mehrdad and Ali Hajbabaie. "On the relationships between commuting mode choice and public health." *Journal of Transport & Health* 4 (2017): 267–77.

Thompson, Allan. "Anthropocentrism: Humanity as peril and promise." In *The Oxford handbook of environmental ethics*, edited by Stephen Gardiner and Allen Thompson, 77–90. New York: Oxford University Press, 2017.

Thornton, Sarah, Selina Pan, Stephen M. Erlien and J. Christian Gerdes. "Incorporating ethical considerations into automated vehicle control." *IEEE Transactions on Intelligent Transportation Systems* 18, no. 6 (2016): 1429–39.

Tillmann, Vera, Meindert Haveman, Reinhilde Stöppler, Štefan Kvas and Daniel Monninger. "Public bus drivers and social inclusion: Evaluation of their knowledge and attitudes toward people with intellectual disabilities." *Journal of Policy and Practice in Intellectual Disabilities* 10, no. 4 (2013): 307–13.

TransitCenter. "Who's onboard 2016." *Transport* 24(4): 284–303. https://transitcenter.org/wp-content/uploads/2016/07/Whos-On-Board-2016-7_12_2016.pdf; transportation_service_because_they_think_uber_will_fill.html.

Uber. "Partnering with transit systems." *Supporting Cities*, 2019. https://www.uber.com/us/en/community/supporting-cities/transit/.

Uber. "What is Uber pass?." https://help.uber.com/riders/article/what-is-uber-pass?nodeId=1f9dbe59-8c13-45f6-8048-689e9c76b3ac.

United States Environmental Protection Agency. "About smog, soot, and other air pollution from transportation." *Smog, Soot, and Other Air Pollution from Transportation*. Last updated March 18, 2019. https://www.epa.gov/transportation-air-pollution-and-climate-change/smog-soot-and-local-air-pollution.

United States Environmental Protection Agency. "Environmental impact assessment." http://www.epa.ie/monitoringassessment/assessment/eia/.

Varzi, Achille. "Mereology." In *The Stanford encyclopedia of philosophy*, edited by E.N. Zalta, 2015. Last modified January 15, 2015. http://plato.stanford.edu/archives/spr2015/entries/mereology/.

Verbeek, Peter-Paul. *Moralizing technology: Understanding and designing the morality of things*. Chicago: University of Chicago Press, 2011.

Wachs, Martin. "Transportation policy, poverty, and sustainability: History and future." *Transportation Research Record* 2163, no. 1 (2010): 5–12.

Walker, Jarrett. "Streetcars: An inconvenient truth." *Human transit: The professional blog of public transit consultant Jarrett Walker*, 2009. http://www.humantransit.org/2009/07/streetcars-an-inconvenienttruth.html.

Walker, Jarrett. "Do Uber and Lyft want to connect to transit?" *Human transit: The professional blog of public transit consultant Jarrett Walker*, 2019. https://humantransit.org/2019/04/do-uber-and-lyft-want-to-connect-to-transit.html.

Walker, Warren and Vincent AWJ Marchau. "Dynamic adaptive policymaking for the sustainable city: The case of automated taxis." *International Journal of Transportation Science and Technology* 6, no. 1 (2017): 1–12.

Watzenig, Daniel and Martin Horn, eds., *Automated driving: Safer and more efficient future driving*. Switzerland: Springer International Publishing, 2016.

Westney, Lynn. "Intrinsic value and the permanent record: The preservation conundrum." *OCLC Systems & Services: International Digital Library Perspectives* (2007): 5–12.

Wild, Kirsty and Alistair Woodward. "Why are cyclists the happiest commuters? Health, pleasure and the e-bike." *Journal of Transport & Health* 14 (2019): 100569.

Index

About the Author

Shane Epting is an assistant professor of philosophy at the Missouri University of Science and Technology, and he is a cofounder of the Philosophy of the City Research Group.

www.ingramcontent.com/pod-product-compliance
Ingram Content Group UK Ltd.
Pitfield, Milton Keynes, MK11 3LW, UK
UKHW041628031026
15587UKWH00001B/129

9 781786 608192